19

P9-DMD-999

THIS IS YOUR BRAIN ON

~MUSIC~

THIS IS YOUR BRAIN ON

~ MUSIC ~

The Science of a Human Obsession

Daniel J. Levitin

DUTTON

DUTTON
Published by Penguin Group (USA) Inc.
375 Hudson Street, New York, New York 10014, U.S.A.
Penguin Group (Canada), 90 Eglinton Avenue East, Suite 700, Toronto, Ontario M4P 2Y3,
Canada (a division of Pearson Penguin Canada Inc.); Penguin Books Ltd, 80 Strand, London
WC2R 0RL, England; Penguin Ireland, 25 St Stephen's Green, Dublin 2, Ireland (a division
of Penguin Books Ltd); Penguin Group (Australia), 250 Camberwell Road, Camberwell,
Victoria 3124, Australia (a division of Pearson Australia Group Pty Ltd); Penguin Books
India Pvt Ltd, 11 Community Centre, Panchsheel Park, New Delhi – 110 017, India; Penguin
Group (NZ), cnr Airborne and Rosedale Roads, Albany, Auckland 1310, New Zealand
(a division of Pearson New Zealand Ltd); Penguin Books (South Africa) (Pty) Ltd,
24 Sturdee Avenue, Rosebank, Johannesburg 2196, South Africa

Penguin Books Ltd, Registered Offices: 80 Strand, London WC2R 0RL, England

Published by Dutton, a member of Penguin Group (USA) Inc.

First printing, August 2006
10 9 8 7 6 5 4 3

REGISTERED TRADEMARK—MARCA REGISTRADA

LIBRARY OF CONGRESS CATALOGING-IN-PUBLICATION DATA

Levitin, Daniel J.
    This is your brain on music : the science of a human obsession / Daniel J. Levitin.
        p.   cm.
    Includes bibliographical references and index.
    ISBN 0-525-94969-0 (hardcover)
    1. Music—Psychological aspects.   I. Title.
    ML3830.L38   2006
    781'.11—dc22   2006009055

Printed in the United States of America
Set in ITC Century Book

CONTENTS

Introduction

I Love Music and I Love Science—Why Would I Want to Mix the Two?

> I love science, and it pains me to think that so many are terrified of the subject or feel that choosing science means you cannot also choose compassion, or the arts, or be awed by nature. Science is not meant to cure us of mystery, but to reinvent and reinvigorate it.
>
> —Robert Sapolsky, *Why Zebras Don't Get Ulcers*, p. xii

In the summer of 1969, when I was eleven, I bought a stereo system at the local hi-fi shop. It cost all of the hundred dollars I had earned weeding neighbors' gardens that spring at seventy-five cents an hour. I spent long afternoons in my room, listening to records: Cream, the Rolling Stones, Chicago, Simon and Garfunkel, Bizet, Tchaikovsky, George Shearing, and the saxophonist Boots Randolph. I didn't listen particularly loud, at least not compared to my college days when I actually set my loudspeakers on fire by cranking up the volume too high, but the noise was evidently too much for my parents. My mother is a novelist; she wrote every day in the den just down the hall and played the piano for an hour every night before dinner. My father was a businessman; he worked eighty-hour weeks, forty of those hours in his office at home on evenings and weekends. Being the businessman that he was, my father made me a proposition: He would buy me a pair of headphones if I would promise to use them when he was home. Those headphones forever changed the way I listened to music.

The new artists that I was listening to were all exploring stereo mixing for the first time. Because the speakers that came with my hundred-dollar all-in-one stereo system weren't very good, I had never before heard the depth that I could hear in the headphones—the placement of instruments both in the left-right field and in the front-back (reverberant) space. To me, records were no longer just about the songs anymore, but about the sound. Headphones opened up a world of sonic colors, a palette of nuances and details that went far beyond the chords and melody, the lyrics, or a particular singer's voice. The swampy Deep South ambience of "Green River" by Creedence, or the pastoral, open-space beauty of the Beatles' "Mother Nature's Son"; the oboes in Beethoven's Sixth (conducted by Karajan), faint and drenched in the atmosphere of a large wood-and-stone church; the sound was an enveloping experience. Headphones also made the music more personal for me; it was suddenly coming from inside my head, not out there in the world. This personal connection is ultimately what drove me to become a recording engineer and producer.

Many years later, Paul Simon told me that the sound is always what he was after too. "The way that I listen to my own records is for the sound of them; not the chords or the lyrics—my first impression is of the overall sound."

I dropped out of college after the incident with the speakers in my dorm room, and I joined a rock band. We got good enough to record at a twenty-four-track studio in California with a talented engineer, Mark Needham, who went on to record hit records by Chris Isaak, Cake, and Fleetwood Mac. Mark took a liking to me, probably because I was the only one interested in going into the control room to hear back what we sounded like, while the others were more interested in getting high in between takes. Mark treated me like a producer, although I didn't know what one was at the time, asking me what the band wanted to sound like. He taught me how much of a difference to the sound a microphone could make, or even the influence of how a microphone was placed. At first, I didn't hear some of the differences he pointed out, but he taught me what to listen for. "Notice that when I put this microphone closer to the guitar amp, the sound becomes fuller, rounder, and more even; but when I put it farther back, it picks up some of the sound of the room, giving it a more spacious sound, although you lose some of the midrange if I do that."

Our band became moderately well known in San Francisco, and our tapes played on local rock radio stations. When the band broke up—due to the guitarist's frequent suicide attempts and the vocalist's nasty habit of taking nitrous oxide and cutting himself with razor blades—I found work as a producer of other bands. I learned to hear things I had never heard before: the difference between one microphone and another, even between one brand of recording tape and another (Ampex 456 tape had a characteristic "bump" in the low-frequency range, Scotch 250 had a characteristic crispness in the high frequencies, and Agfa 467 a luster in the midrange). Once I knew what to listen for, I could tell Ampex from Scotch or Agfa tape as easily as I could tell an apple from a pear or an orange. I progressed to work with other great engineers, like Leslie Ann Jones (who had worked with Frank Sinatra and Bobby McFerrin), Fred Catero (Chicago, Janis Joplin), and Jeffrey Norman (John Fogerty, the Grateful Dead). Even though I was the producer—the person in charge of the sessions—I was intimidated by them all. Some of the engineers let me sit in on their sessions with other artists, such as Heart, Journey, Santana, Whitney Houston, and Aretha Franklin. I got a lifetime of education watching them interact with the artists, talking about subtle nuances in how a guitar part was articulated or how a vocal performance had been delivered. They would talk about syllables in a lyric, and choose among ten different performances. They could hear so well; how did they train their ears to hear things that mere mortals couldn't?

While working with small, unknown bands, I got to know the studio managers and engineers, and they steered me toward better and better work. One day an engineer didn't show up and I spliced some tape edits for Carlos Santana. Another time, the great producer Sandy Pearlman went out for lunch during a Blue Öyster Cult session and left me in charge to finish the vocals. One thing led to another, and I spent over a decade producing records in California; I was eventually lucky enough to be able to work with many well-known musicians. But I also worked with dozens of musical no-names, people who are extremely talented but never made it. I began to wonder why some musicians become household names while others languish in obscurity. I also wondered why music seemed to come so easily to some and not others. Where does creativity come from? Why do some songs move us so and others leave us cold? And what about the role of perception in all of this, the uncanny ability of great musicians and engineers to hear nuances that most of us don't?

These questions led me back to school for some answers. While still working as a record producer, I drove down to Stanford University twice a week with Sandy Pearlman to sit in on neuropsychology lectures by Karl Pribram. I found that psychology was the field that held the answers to some of my questions—questions about memory, perception, creativity, and the common instrument underlying all of these: the human brain. But instead of finding answers, I came away with more questions—as is often the case in science. Each new question opened my mind to an appreciation for the complexity of music, of the world, and of the human experience. As the philosopher Paul Churchland notes, humans have been trying to understand the world throughout most of recorded history; in just the past two hundred years, our curiosity has revealed much of what Nature had kept hidden from us: the fabric of space-time, the constitution of matter, the many forms of energy, the origins of the universe, the nature of life itself with the discovery of DNA, and the completion of the mapping of the human genome just five years ago. But one mystery has not been solved: the mystery of the human brain and how it gives rise to thoughts and feelings, hopes and desires, love, and the experience of beauty, not to mention dance, visual art, literature, and music.

What is music? Where does it come from? Why do some sequences of sounds move us so, while others—such as dogs barking or cars screeching—make many people uncomfortable? For some of us, these questions occupy a large part of our life's work. For others, the idea of picking music apart in this way seems tantamount to studying the chemical structure in a Goya canvas, at the expense of seeing the art that the painter was trying to produce. The Oxford historian Martin Kemp points out a similarity between artists and scientists. Most artists describe their work as experiments—part of a series of efforts designed to explore a common concern or to establish a viewpoint. My good friend and colleague William Forde Thompson (a music cognition scientist and composer at the University of Toronto) adds that the work of both scientists and artists involves similar stages of development: a creative and exploratory "brainstorming" stage, followed by testing and refining stages that typically involve the application of set procedures, but are often informed by additional creative problem-solving. Artists' studios and scientists' laboratories share similarities as well, with a large number of projects going at once, in various stages of incompletion. Both require specialized tools, and the results are—unlike the final plans for a suspension bridge, or the tallying of money in a bank account at the end of the business day—open to interpretation. What artists and scientists have in common is the ability to live in an open-ended state of interpretation and reinterpretation of the products of our work. The work of artists and scientists is ultimately the pursuit of truth, but members of both camps understand that truth in its very nature is contextual and changeable, dependent on point of view, and that today's truths become tomorrow's disproven hypotheses or forgotten objets d'art. One need look no further than Piaget, Freud, and Skinner to find theories that once held widespread currency and were later overturned (or at least dramatically reevaluated). In music, a number of groups were prematurely held up as of lasting importance: Cheap Trick were hailed as the new Beatles, and at one time the *Rolling Stone Encyclopedia of Rock* devoted as much space to Adam and the Ants as they did to U2. There were times when people couldn't imagine a day when most of the world would not know the names Paul Stookey, Christopher Cross, or Mary Ford. For the artist, the goal of the painting or musical composition is not to convey literal truth, but an aspect of a universal truth that if successful, will continue to move and to touch people even as contexts, societies, and cultures change. For the scientist, the goal of a theory is to convey "truth for now"—to replace an old truth, while accepting that someday this theory, too, will be replaced by a new "truth," because that is the way science advances.

Music is unusual among all human activities for both its *ubiquity* and its *antiquity*. No known human culture now or anytime in the recorded past lacked music. Some of the oldest physical artifacts found in human and protohuman excavation sites are musical instruments: bone flutes and animal skins stretched over tree stumps to make drums. Whenever humans come together for any reason, music is there: weddings, funerals, graduation from college, men marching off to war, stadium sporting events, a night on the town, prayer, a romantic dinner, mothers rocking their infants to sleep, and college students studying with music as a background. Even more so in nonindustrialized cultures than in modern Western societies, music is and was part of the fabric of everyday life. Only relatively recently in our own culture, five hundred years or so ago, did a distinction arise that cut society in two, forming separate classes of music performers and music listeners. Throughout most of the world and for most of human history, music making was as natural an activity as breathing and walking, and everyone participated. Concert halls, dedicated to the performance of music, arose only in the last several centuries.

Jim Ferguson, whom I have known since high school, is now a professor of anthropology. Jim is one of the funniest and most fiercely intelligent people I know, but he is shy—I don't know how he manages to teach his lecture courses. For his doctoral degree at Harvard, he performed fieldwork in Lesotho, a small nation completely surrounded by South Africa. There, studying and interacting with local villagers, Jim patiently earned their trust until one day he was asked to join in one of their songs. So, typically, when asked to sing with these Sotho villagers, Jim said in a soft voice, "I don't sing," and it was true: We had been in high school band together and although he was an excellent oboe player, he couldn't carry a tune in a bucket. The villagers found his objection puzzling and inexplicable. The Sotho consider singing an ordinary, everyday activity performed by everyone, young and old, men and women, not an activity reserved for a special few.

Our culture, and indeed our very language, makes a distinction between a class of expert performers—the Arthur Rubinsteins, Ella Fitzgeralds, Paul McCartneys—and the rest of us. The rest of us pay money to hear the experts entertain us. Jim knew that he wasn't much of a singer or dancer, and to him, a public display of singing and dancing implied he thought himself an expert. The villagers just stared at Jim and said, "What do you mean you don't sing?! You talk!" Jim told me later, "It was as odd to them as if I told them that I couldn't walk or dance, even though I have both my legs." Singing and dancing were a natural activity in everybody's lives, seamlessly integrated and involving everyone. The Sesotho verb for singing *(ho bina)*, as in many of the world's languages, also means to dance; there is no distinction, since it is assumed that singing involves bodily movement.

A couple of generations ago, before television, many families would sit around and play music together for entertainment. Nowadays there is a great emphasis on technique and skill, and whether a musician is "good enough" to play for others. Music making has become a somewhat reserved activity in our culture, and the rest of us listen. The music industry is one of the largest in the United States, employing hundreds of thousands of people. Album sales alone bring in $30 billion a year, and this figure doesn't even account for concert ticket sales, the thousands of bands playing Friday nights at saloons all over North America, or the thirty billion songs that were downloaded free through peer-to-peer file sharing in 2005. Americans spend more money on music than on sex or prescription drugs. Given this voracious consumption, I would say that most Americans qualify as expert music listeners. We have the cognitive capacity to detect wrong notes, to find music we enjoy, to remember hundreds of melodies, and to tap our feet in time with the music—an activity that involves a process of meter extraction so complicated that most computers cannot do it. Why do we listen to music, and why are we willing to spend so much money on music listening? Two concert tickets can easily cost as much as a week's food allowance for a family of four, and one CD costs about the same as a work shirt, eight loaves of bread, or basic phone service for a month. Understanding why we like music and what draws us to it is a window on the essence of human nature.

To ask questions about a basic, and omnipresent human ability is to implicitly ask questions about evolution. Animals evolved certain physical forms as a response to their environment, and the characteristics that conferred an advantage for mating were passed down to the next generation through the genes.

A subtle point in Darwinian theory is that living organisms—whether plants, viruses, insects, or animals—coevolved with the physical world. In other words, while all living things are changing in response to the world, the world is also changing in response to them. If one species develops a mechanism to keep away a particular predator, that predator's species is then under evolutionary pressure either to develop a means to overcome that defense or to find another food source. Natural selection is an arms race of physical morphologies changing to catch up with one another.

A relatively new scientific field, evolutionary psychology, extends the notion of evolution from the physical to the realm of the mental. My mentor when I was a student at Stanford University, the cognitive psychologist Roger Shepard, notes that not just our bodies but our minds are the product of millions of years of evolution. Our thought patterns, our predispositions to solve problems in certain ways, our sensory systems—such as the ability to see color (and the particular colors we see)—are all products of evolution. Shepard pushes the point still further: Our minds coevolved with the physical world, changing in response to ever-changing conditions. Three of Shepard's students, Leda Cosmides and John Tooby of the University of California at Santa Barbara, and Geoffrey Miller of the University of New Mexico, are among those at the forefront of this new field. Researchers in this field believe that they can learn a lot about human behavior by considering the evolution of the mind. What function did music serve humankind as we were evolving and developing? Certainly the music of fifty thousand and one hundred thousand years ago is very different from Beethoven, Van Halen, or Eminem. As our brains have evolved, so has the music we make with them, and the music that we want to hear. Did particular regions and pathways evolve in our brains specifically for making and listening to music?

Contrary to the old, simplistic notion that art and music are processed in the right hemisphere of our brains, with language and mathematics in the left, recent findings from my laboratory and those of my colleagues are showing us that music is distributed throughout the brain. Through studies of people with brain damage, we've seen patients who have lost the ability to read a newspaper but can still read music, or individuals who can play the piano but lack the motor coordination to button their own sweater. Music listening, performance, and composition engage nearly every area of the brain that we have so far identified, and involve nearly every neural subsystem. Could this fact account for claims that music listening exercises other parts of our minds; that listening to Mozart twenty minutes a day will make us smarter?

The power of music to evoke emotions is harnessed by advertising executives, filmmakers, military commanders, and mothers. Advertisers use music to make a soft drink, beer, running shoe, or car seem more hip than their competitors'. Film directors use music to tell us how to feel about scenes that otherwise might be ambiguous, or to augment our feelings at particularly dramatic moments. Think of a typical chase scene in an action film, or the music that might accompany a lone woman climbing a staircase in a dark old mansion: Music is being used to manipulate our emotions, and we tend to accept, if not outright enjoy, the power of music to make us experience these different feelings. Mothers throughout the world, and as far back in time as we can imagine, have used soft singing to soothe their babies to sleep, or to distract them from something that has made them cry.

Many people who love music profess to know nothing about it. I've found that many of my colleagues who study difficult, intricate topics such as neurochemistry or psychopharmacology feel unprepared to deal with research in the neuroscience of music. And who can blame them? Music theorists have an arcane, rarified set of terms and rules that are as obscure as some of the most esoteric domains of mathematics. To the nonmusician, the blobs of ink on a page that we call music notation might just as well be the notations of mathematical set theory. Talk of keys, cadences, modulation, and transposition can be baffling.

Yet every one of my colleagues who feels intimidated by such jargon can tell me the music that he or she likes. My friend Norman White is a world authority on the hippocampus in rats, and how they remember different places they've visited. He is a huge jazz fan, and can talk expertly about his favorite artists. He can instantly tell the difference between Duke Ellington and Count Basie by the sound of the music, and can even tell early Louis Armstrong from late. Norm doesn't have any knowledge about music in the technical sense—he can tell me that he likes a certain song, but he can't tell me what the names of the chords are. He is, however, an expert in knowing what he likes. This is not at all unusual, of course. Many of us have a practical knowledge of things we like, and can communicate our preferences without possessing the technical knowledge of the true expert. I know that I prefer the chocolate cake at one restaurant I often go to, over the chocolate cake at my neighborhood coffee shop. But only a chef would be able to analyze the cake—to decompose the taste experience into its elements—by describing the differences in the kind of flour, or the shortening, or the type of chocolate used.

It's a shame that many people are intimidated by the jargon musicians, music theorists, and cognitive scientists throw around. There is specialized vocabulary in every field of inquiry (try to make sense of a full blood-analysis report from your doctor). But in the case of music, music experts and scientists could do a better job of making their work accessible. That is something I tried to accomplish in this book. The unnatural gap that has grown between musical performance and music listening has been paralleled by a gap between those who love music (and love to talk about it) and those who are discovering new things about how it works.

A feeling my students often confide to me is that they love life and its mysteries, and they're afraid that too much education will steal away many of life's simple pleasures. Robert Sapolsky's students have probably confided much the same to him, and I myself felt the same anxiety in 1979, when I moved to Boston to attend the Berklee College of Music. What if I took a scholarly approach to studying music and, in analyzing it, stripped it of its mysteries? What if I became so knowledgeable about music that I no longer took pleasure from it?

I still take as much pleasure from music as I did from that cheap hi-fi through those headphones. The more I learned about music and about science the more fascinating they became, and the more I was able to appreciate people who are really good at them. Like science, music over the years has proved to be an adventure, never experienced exactly the same way twice. It has been a source of continual surprise and satisfaction for me. It turns out science and music aren't such a bad mix.

This book is about the science of music, from the perspective of cognitive neuroscience—the field that is at the intersection of psychology and neurology. I'll discuss some of the latest studies I and other researchers in our field have conducted on music, musical meaning, and musical pleasure. They offer new insights into profound questions. If all of us hear music differently, how can we account for pieces that seem to move so many people—Handel's *Messiah* or Don McLean's "Vincent (Starry Starry Night)" for example? On the other hand, if we all hear music in the same way, how can we account for wide differences in musical preference—why is it that one man's Mozart is another man's Madonna?

The mind has been opened up in the last few years by the exploding field of neuroscience and the new approaches in psychology due to new brain-imaging technologies, drugs able to manipulate neurotransmitters such as dopamine and serotonin, and plain old scientific pursuit. Less well known are the extraordinary advances we have been able to make in modeling how our neurons network, thanks to the continuing revolution in computer technology. We are coming to understand computational systems in our head like never before. Language now seems to be substantially hardwired into our brains. Even consciousness itself is no longer hopelessly shrouded in a mystical fog, but is rather something that emerges from observable physical systems. But no one until now has taken all this new work together and used it to elucidate what is for me the most beautiful human obsession. Your brain on music is a way to understand the deepest mysteries of human nature. That is why I wrote this book.

By better understanding what music is and where it comes from, we may be able to better understand our motives, fears, desires, memories, and even communication in the broadest sense. Is music listening more along the lines of eating when you're hungry, and thus satisfying an urge? Or is it more like seeing a beautiful sunset or getting a backrub, which triggers sensory pleasure systems in the brain? Why do people seem to get stuck in their musical tastes as they grow older and cease experimenting with new music? This is the story of how brains and music co-evolved—what music can teach us about the brain, what the brain can teach us about music, and what both can teach us about ourselves.

1. What Is Music?

From Pitch to Timbre

What is music? To many, "music" can only mean the great masters—Beethoven, Debussy, and Mozart. To others, "music" is Busta Rhymes, Dr. Dre, and Moby. To one of my saxophone teachers at Berklee College of Music—and to legions of "traditional jazz" aficionados—anything made before 1940 or after 1960 isn't really music at all. I had friends when I was a kid in the sixties who used to come over to my house to listen to the Monkees because their parents forbade them to listen to anything but classical music, and others whose parents would only let them listen to and sing religious hymns. When Bob Dylan dared to play an electric guitar at the Newport Folk Festival in 1965, people walked out and many of those who stayed, booed. The Catholic Church banned music that contained polyphony (more than one musical part playing at a time), fearing that it would cause people to doubt the unity of God. The church also banned the musical interval of an augmented fourth, the distance between C and F-sharp and also known as a tritone (the interval in Leonard Bernstein's *West Side Story* when Tony sings the name "Maria"). This interval was considered so dissonant that it must have been the work of Lucifer, and so the church named it *Diabolus in musica*. It was pitch that had the medieval church in an uproar. And it was timbre that got Dylan booed.

The music of avant-garde composers such as Francis Dhomont, Robert Normandeau, or Pierre Schaeffer stretches the bounds of what most of us think music is. Going beyond the use of melody and harmony, and even beyond the use of instruments, these composers use recordings of found objects in the world such as jackhammers, trains, and waterfalls. They edit the recordings, play with their pitch, and ultimately combine them into an organized collage of sound with the same type of emotional trajectory—the same tension and release—as traditional music. Composers in this tradition are like the painters who stepped outside of the boundaries of representational and realistic art—the cubists, the Dadaists, many of the modern painters from Picasso to Kandinsky to Mondrian.

What do the music of Bach, Depeche Mode, and John Cage fundamentally have in common? On the most basic level, what distinguishes Busta Rhymes's "What's It Gonna Be?!" or Beethoven's "Pathétique" Sonata from, say, the collection of sounds you'd hear standing in the middle of Times Square, or those you'd hear deep in a rainforest? As the composer Edgard Varèse famously defined it, "Music is organized sound."

This book drives at a neuropsychological perspective on how music affects our brains, our minds, our thoughts, and our spirit. But first, it is helpful to examine what music is made of. What are the fundamental building blocks of music? And how, when organized, do they give rise to music? The basic elements of any sound are loudness, pitch, contour, duration (or rhythm), tempo, timbre, spatial location, and reverberation. Our brains organize these fundamental perceptual attributes into higher-level concepts—just as a painter arranges lines into forms—and these include meter, harmony, and melody. When we listen to music, we are actually perceiving multiple attributes or "dimensions." Here is a brief summary of them.

~ A discrete musical sound is usually called a tone. The word *note* is also used, but scientists reserve that word to refer to something that is notated on a page or score of music. The two terms, *tone* and *note*, refer to the same entity in the abstract, where the word *tone* refers to what you hear, and the word *note* refers to what you see written on a musical score.

~ Pitch is a purely psychological construct, related both to the actual frequency of a particular tone and to its relative position in the musical scale. It provides the answer to the question "What note is that?" ("It's a C-sharp.") I'll define frequency and musical scale below.

~ Rhythm refers to the durations of a series of notes, and to the way that they group together into units. For example, in the "Alphabet Song" (the same as "Twinkle, Twinkle Little Star") the notes of the song are all equal in duration for the letters A B C D E F G H I J K (with an equal duration pause, or rest, between G and H), and then the following four letters are sung with half the duration, or twice as fast per letter: L M N O (leading generations of schoolchildren to spend several early months believing that there was a letter in the English alphabet called ellemmenno).

~ Tempo refers to the overall speed or pace of the piece.

~ Contour describes the overall shape of a melody, taking into account only the pattern of "up" and "down" (whether a note goes up or down, not the amount by which it goes up or down).

~ Timbre is that which distinguishes one instrument from another—say, trumpet from piano—when both are playing the same written note. It is a kind of tonal color that is produced in part by overtones from the instrument's vibrations.

~ Loudness is a purely psychological construct that relates (nonlinearly and in poorly understood ways) to the physical amplitude of a tone.

~ Spatial location is where the sound is coming from.

~ Reverberation refers to the perception of how distant the source is from us in combination with how large a room or hall the music is in; often referred to as "echo" by laypeople, it is the quality that distinguishes the spaciousness of singing in a large concert hall from the sound of singing in your shower. It has an underappreciated role in communicating emotion and creating an overall pleasing sound.

These attributes are separable. Each can be varied without altering the others, allowing the scientific study of one at a time, which is why we can think of them as dimensions. The difference between *music* and a random or disordered set of sounds has to do with the way these fundamental attributes combine, and the relations that form between them. When these basic elements combine and form relationships with one another in a meaningful way, they give rise to higher-order concepts such as meter, key, melody, and harmony.

~ Meter is created by our brains by extracting information from rhythm and loudness cues, and refers to the way in which tones are grouped with one another across time. A waltz meter organizes tones into groups of three, a march into groups of two or four.

~ Key has to do with a hierarchy of importance that exists between tones in a musical piece; this hierarchy does not exist in-the-world, it exists only in our minds, as a function of our experiences with a musical style and musical idioms, and mental schemas that all of us develop for understanding music.

~ Melody is the main theme of a musical piece, the part you sing along with, the succession of tones that are most salient in your mind. The notion of melody is different across genres. In rock music, there is typically a melody for the verses and a melody for the chorus, and verses are distinguished by a change in lyrics and sometimes by a change in instrumentation. In classical music, the melody is a starting point for the composer to create variations on that theme, which may be used throughout the entire piece in different forms.

~ Harmony has to do with relationships between the pitches of different tones, and with tonal contexts that these pitches set up that ultimately lead to expectations for what will come next in a musical piece—expectations that a skillful composer can either meet or violate for artistic and expressive purposes. Harmony can mean simply a parallel melody to the primary one (as when two singers harmonize) or it can refer to a chord progression—the clusters of notes that form a context and background on which the melody rests.

The idea of primitive elements combining to create art, and of the importance of relationships between elements, also exists in visual art and dance. The fundamental elements of visual perception include color (which can be decomposed into the three dimensions of hue, saturation, and lightness), brightness, location, texture, and shape. But a painting is more than these—it is not just a line here and another there, or a spot of red in one part of the picture and a patch of blue in another. What makes a set of lines and colors into art is the *relationship* between this line and that one; the way one color or form echoes another in a different part of the canvas. Those dabs of paint and lines become art when form and flow (the way in which your eye is drawn across the canvas) are created out of lower-level perceptual elements. When they combine harmoniously they ultimately give rise to perspective, foreground and background, emotion, and other aesthetic attributes. Similarly, dance is not just a raging sea of unrelated bodily movements; the relationship of those movements to one another is what creates integrity and integrality, a coherence and cohesion that the higher levels of our brain process. And as in visual art, music plays on not just what notes are sounded, but which ones are not. Miles Davis famously described his improvisational technique as parallel to the way that Picasso described his use of a canvas: The most critical aspect of the work, both artists said, was not the objects themselves, but the space between objects. In Miles's case, he described the most important part of his solos as the empty space between notes, the "air" that he placed between one note and the next. Knowing precisely when to hit the next note, and allowing the listener time to anticipate it, is a hallmark of Davis's genius. This is particularly apparent in his album *Kind of Blue*.

To nonmusicians, terms such as *diatonic*, *cadence*, or even *key* and *pitch* can throw up an unnecessary barrier. Musicians and critics sometimes appear to live behind a veil of technical terms that can sound pretentious. How many times have you read a concert review in the newspaper and found you have no idea what the reviewer is saying? "Her sustained *appoggiatura* was flawed by an inability to complete the *roulade*." Or, "I can't believe they modulated to C-sharp minor! How ridiculous!" What we really want to know is whether the music was performed in a way that moved the audience. Whether the singer seemed to inhabit the character she was singing about. You might want the reviewer to compare tonight's performance to that of a previous night or a different ensemble. We're usually interested in the music, not the technical devices that were used. We wouldn't stand for it if a restaurant reviewer started to speculate about the precise temperature at which the chef introduced the lemon juice in a hollandaise sauce, or if a film critic talked about the aperture of the lens that the cinematographer used; we shouldn't stand for it in music either.

Moreover, many of those who study music—even musicologists and scientists—disagree about what is meant by some of these terms. We employ the term *timbre*, for example, to refer to the overall sound or tonal color of an instrument—that indescribable character that distinguishes a trumpet from a clarinet when they're playing the same written note, or what distinguishes your voice from Brad Pitt's if you're saying the same words. But an inability to agree on a definition has caused the scientific community to take the unusual step of throwing up its hands and defining timbre by what it is not. (The official definition of the Acoustical Society of America is that timbre is everything about a sound that is not loudness or pitch. So much for scientific precision!)

What is pitch? This simple question has generated hundreds of scientific articles and thousands of experiments. Pitch is related to the frequency or rate of vibration of a string, column of air, or other physical source. If a string is vibrating so that it moves back and forth sixty times in one second, we say that it has a frequency of sixty cycles per second. The unit of measurement, cycles per second, is often called Hertz (abbreviated Hz) after Heinrich Hertz, the German theoretical physicist who was the first to transmit radio waves (a dyed-in-the-wool theoretician, when asked what practical use radio waves might have, he reportedly shrugged, "None"). If you were to try to mimic the sound of a fire engine siren, your voice would sweep through different pitches, or frequencies (as the tension in your vocal folds changes), some "low" and some "high."

Keys on the left of the piano keyboard strike longer, thicker strings that vibrate at a relatively slow rate. Keys to the right strike shorter, thinner strings that vibrate at a higher rate. The vibration of these strings displaces air molecules, and causes them to vibrate at the same rate—with the same frequency as the string. These vibrating air molecules are what reach our eardrum, and they cause our eardrum to wiggle in and out at the same frequency. The only information that our brains get about the pitch of sound comes from that wiggling in and out of our eardrum; our inner ear and our brain have to analyze the motion of the eardrum in order to figure out what vibrations out-there-in-the-world caused the eardrum to move that way.

By convention, when we press keys nearer to the left of the keyboard, we say that they are "low" pitch sounds, and ones near the right side of the keyboard are "high" pitch. That is, what we call "low" are those sounds that vibrate slowly, and are closer (in vibration frequency) to the sound of a large dog barking. What we call "high" are those sounds that vibrate rapidly, and are closer to what a small yip-yip dog might make. But even these terms *high* and *low* are culturally relative—the Greeks talked about sounds in the opposite way because the stringed instruments they built tended to be oriented vertically. Shorter strings or pipe organ tubes had their tops closer to the ground, so these were called the "low" notes (as in "low to the ground,") and the longer strings and tubes—reaching up toward Zeus and Apollo—were called the "high" notes. *Low* and *high*—just like *left* and *right*—are effectively arbitrary terms that ultimately have to be memorized. Some writers have argued that "high" and "low" are intuitive labels, noting that what we call high-pitched sounds come from birds (who are high up in trees or in the sky) and what we call low-pitched sounds often come from large, close-to-the-ground mammals such as bears or the low sounds of an earthquake. But this is not convincing, since low sounds also come from up high (think of thunder) and high sounds can come from down low (crickets and squirrels, leaves being crushed underfoot).

As a first definition of *pitch*, let's say it is that quality that primarily distinguishes the sound that is associated with pressing one piano key versus another.

Pressing a piano key causes a hammer to strike one or more strings inside the piano. Striking a string displaces it, stretching it a bit, and its inherent resiliency causes it to return toward its original position. But it overshoots that original position, going too far in the opposite direction, and then attempts to return to its original position again, overshooting it again, and in this way it oscillates back and forth. Each oscillation covers less distance, and, in time, the string stops moving altogether. This is why the sound you hear when you press a piano key gets softer until it trails off into nothing. The distance that the string covers with each oscillation back and forth is translated by our brains into loudness; the rate at which it oscillates is translated into pitch. The farther the string travels, the louder the sound seems to us; when it is barely traveling at all, the sound seems soft. Although it might seem counterintuitive, the distance traveled and the rate of oscillation are independent. A string can vibrate very quickly and traverse either a great distance or a small one. The distance it traverses is related to how hard we hit it—this corresponds to our intuition that hitting something harder makes a louder sound. The rate at which the string vibrates is principally affected by its size and how tightly strung it is, not by how hard it was struck.

It might seem as though we should simply say that pitch is the same as frequency; that is, the frequency of vibration of air molecules. This is almost true. Mapping the physical world onto the mental world is seldom so straightforward. However, for most musical sounds, pitch and frequency are closely related.

The word pitch refers to the mental representation an organism has of the fundamental frequency of a sound. That is, pitch is a purely psychological phenomenon related to the frequency of vibrating air molecules. By "psychological," I mean that it is entirely in our heads, not in the world-out-there; it is the end product of a chain of mental events that gives rise to an entirely subjective, internal mental representation or quality. Sound waves—molecules of air vibrating at various frequencies—do not themselves have pitch. Their motion and oscillations can be measured, but it takes a human (or animal) brain to map them to that internal quality we call pitch.

We perceive color in a similar way, and it was Isaac Newton who first realized this. (Newton, of course, is known as the discoverer of the theory of gravity, and the inventor, along with Leibniz, of calculus. Like Einstein, Newton was a very poor student, and his teachers often complained of his inattentiveness. Ultimately, Newton was kicked out of school.)

Newton was the first to point out that light is colorless, and that consequently color has to occur inside our brains. He wrote, "The waves themselves are not colored." Since his time, we have learned that light waves are characterized by different frequencies of oscillation, and when they impinge on the retina of an observer, they set off a chain of neurochemical events, the end product of which is an internal mental image that we call color. The essential point here is: What we perceive as color is not made up of color. Although an apple may appear red, its atoms are not themselves red. And similarly, as the philosopher Daniel Dennett points out, heat is not made up of tiny hot things.

A bowl of pudding only has taste when I put it in my mouth—when it is in contact with my tongue. It doesn't have taste or flavor sitting in my fridge, only the potential. Similarly, the walls in my kitchen are not "white" when I leave the room. They still have paint on them, of course, but *color* only occurs when they interact with my eyes.

Sound waves impinge on the eardrums and pinnae (the fleshy parts of your ear), setting off a chain of mechanical and neurochemical events, the end product of which is an internal mental image we call pitch. If a tree falls in a forest and no one is there to hear it, does it make a sound? (The question was first posed by the Irish philosopher George Berkeley.) Simply, no—sound is a mental image created by the brain in response to vibrating molecules. Similarly, there can be no pitch without a human or animal present. A suitable measuring device can register the frequency made by the tree falling, but truly it is not pitch unless and until it is heard.

No animal can hear a pitch for every frequency that exists, just as the colors that we actually see are a small portion of the entire electromagnetic spectrum. Sound can theoretically be heard for vibrations from just over 0 cycles per second up to 100,000 cycles per second or more, but each animal hears only a subset of the possible sounds. Humans who are not suffering from any kind of hearing loss can usually hear sounds from 20 Hz to 20,000 Hz. The pitches at the low end sound like an indistinct rumble or shaking—this is the sound we hear when a truck goes by outside the window (its engine is creating sound around 20 Hz) or when a tricked-out car with a fancy sound system has the subwoofers cranked up really loud. Some frequencies—those below 20 Hz—are inaudible to humans because the physiological properties of our ears aren't sensitive to them.

The range of human hearing is generally 20 Hz to 20,000 Hz, but this doesn't mean that the range of human pitch perception is the same; although we can hear sounds in this entire range, they don't all sound musical; that is, we can't unambiguously assign a pitch to the entire range. By analogy, colors at the infrared and ultraviolet ends of the spectrum lack definition compared to the colors closer to the middle. The figure on page 23 shows the range of musical instruments, and the frequency associated with them. The sound of the average male speaking voice is around 110 Hz, and the average female speaking voice is around 220 Hz. The hum of fluorescent lights or from faulty wiring is 60 Hz (in North America; in Europe and countries with a different voltage/current stan-

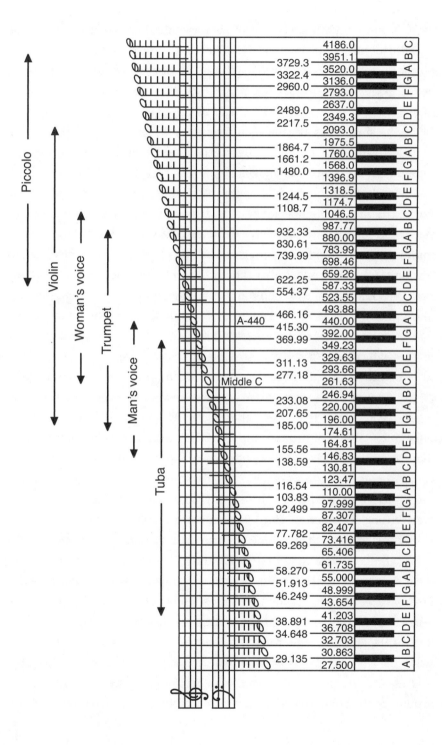

dard, it can be 50 Hz). The sound that a singer hits when she causes a glass to break might be 1000 Hz. The glass breaks because it, like all physical objects, has a natural and inherent vibration frequency. You can hear this by flicking your finger against its sides or, if it's crystal, by running your wet finger around the rim of the glass in a circular motion. When the singer hits just the right frequency—the resonant frequency of the glass—it causes the molecules of the glass to vibrate at their natural rate, and they vibrate themselves apart.

A standard piano has eighty-eight keys. Very rarely, pianos can have a few extra ones at the bottom and electronic pianos, organs, and synthesizers can have as few as twelve or twenty-four keys, but these are special cases. The lowest note on a standard piano vibrates with a frequency of 27.5 Hz. Interestingly, this is about the same rate of motion that constitutes an important threshold in visual perception. A sequence of still photographs—slides—displayed at or about this rate of presentation will give the illusion of motion. "Motion pictures" are a sequence of still images alternating with pieces of black film presented at a rate (one forty-eighth of a second) that exceeds the temporal resolving properties of the human visual system. We perceive smooth, continuous motion when in fact there is no such thing actually being shown to us. When molecules vibrate at around this speed we hear something that sounds like a continuous tone. If you put playing cards in the spokes of your bicycle wheel when you were a kid, you demonstrated to yourself a related principle: At slow speeds, you simply hear the click-click-click of the card hitting the spokes. But above a certain speed, the clicks run together and create a buzz, a tone you can actually hum along with; a pitch.

When this lowest note on the piano plays, and vibrates at 27.5 Hz, to most people it lacks the distinct pitch of sounds toward the middle of the keyboard. At the lowest and the highest ends of the piano keyboard, the notes sound fuzzy to many people with respect to their pitch. Composers know this, and they either use these notes or avoid them depending on what they are trying to accomplish compositionally and emotionally. Sounds with frequencies above the highest note on the piano keyboard, around 6000 Hz and more, sound like a high-pitched whistling to most people. Above 20,000 Hz most humans don't hear a thing, and by the age of sixty, most adults can't hear much above 15,000 Hz or so due to a stiffening of the hair cells in the inner ear. So when we talk about the range of musical notes, or that restricted part of the piano keyboard that conveys the strongest sense of pitch, we are talking about roughly three quarters of the notes on the piano keyboard, between about 55 Hz and 2000 Hz.

Pitch is one of the primary means by which musical emotion is conveyed. Mood, excitement, calm, romance, and danger are signaled by a number of factors, but pitch is among the most decisive. A single high note can convey excitement, a single low note sadness. When notes are strung together, we get more powerful and more nuanced musical statements. Melodies are defined by the pattern or relation of successive pitches across time; most people have no trouble recognizing a melody that is played in a higher or lower key than they've heard it in before. In fact, many melodies do not have a "correct" starting pitch, they just float freely in space, starting anywhere. "Happy Birthday" is an example of this. One way to think about a melody, then, is as an abstract prototype that is derived from specific combinations of key, tempo, instrumentation, and so on. A melody is an auditory object that maintains its identity in spite of transformations, just as a chair maintains its identity when you move it to the other side of the room, turn it upside down, or paint it red. So, for example, if you hear a song played louder than you are accustomed to, you still identify it as the same song. The same holds for changes in the absolute pitch values of the song, which can be changed so long as the relative distances between them remain the same.

The notion of relative pitch values is seen readily in the way that we speak. When you ask someone a question, your voice naturally rises in intonation at the end of the sentence, signaling that you are asking. But you don't try to make the rise in your voice match a specific pitch. It is enough that you end the sentence somewhat higher in pitch than you began it. This is a convention in English (though not in all languages—we have to learn it), and is known in linguistics as a prosodic cue. There are similar conventions for music written in the Western tradition. Certain sequences of pitches evoke calm, others, excitement. The brain basis for this is primarily based on learning, just as we learn that a rising intonation indicates a question. All of us have the innate capacity to learn the linguistic and musical distinctions of whatever culture we are born into, and experience with the music of that culture shapes our neural pathways so that we ultimately internalize a set of rules common to that musical tradition.

Different instruments use different parts of the range of available pitches. The piano has the largest range of any instrument, as you can see from the previous illustration. The other instruments each use a subset of the available pitches, and this influences the ways that instruments are used to communicate emotion. The piccolo, with its high-pitched, shrill, and birdlike sound, tends to evoke flighty, happy moods regardless of the notes it's playing. Because of this, composers tend to use the piccolo for happy music, or rousing music, as in a Sousa march. Similarly, in *Peter and the Wolf*, Prokofiev uses the flute to represent the bird, and the French horn to indicate the wolf. The characters' individuality in *Peter and the Wolf* is expressed in the timbres of different instruments and each has a leitmotiv—an associated melodic phrase or figure that accompanies the reappearance of an idea, person, or situation. (This is especially true of Wagnerian music drama.) A composer who picks so-called sad pitch sequences would only give these to the piccolo if he were trying to be ironic. The lumbering, deep sounds of the tuba or double bass are often used to evoke solemnity, gravity, or weight.

How many unique pitches are there? Because pitch comes from a continuum—the vibration frequencies of molecules—there are technically an infinite number of pitches: For every pair of frequencies you mention, I could always come up with one between them, and a theoretically different pitch would exist. But not every change in frequency gives rise to a noticeable difference in pitch, just as adding a grain of sand to your backpack will not change the weight perceptibly. So not all frequency changes are musically useful. People differ in their ability to detect small changes in frequency; training can help, but generally speaking, most cultures don't use distances much smaller than a semitone as the basis for their music, and most people can't reliably detect changes smaller than about one tenth of a semitone.

The ability to detect differences in pitch is based on physiology, and varies from one animal to another. The basilar membrane of the human inner ear contains hair cells that are frequency selective, firing only in response to a certain band of frequencies. These are stretched out across the membrane from low frequencies to high; low-frequency sounds excite hair cells on one end of the basilar membrane, medium frequency sounds excite the hair cells in the middle, and high-frequency sounds excite them at the other end. We can think of the membrane as containing a map of different pitches very much like a piano keyboard superimposed on it. Because the different tones are spread out across the surface topography of the membrane, this is called a tonotopic map.

After sounds enter the ear, they pass by the basilar membrane, where certain hair cells fire, depending on the frequency of the sounds. The membrane acts like a motion-detector lamp you might have in your garden; activity in a certain part of the membrane causes it to send an electrical signal on up to the auditory cortex. The auditory cortex also has a tonotopic map, with low to high tones stretched out across the cortical surface. In this sense, the brain contains a "map" of different pitches, and different areas of the brain respond to different pitches. Pitch is so important that the brain represents it directly; unlike almost any other musical attribute, we could place electrodes in the brain and be able to determine what pitches were being played to a person just by looking at the brain activity. And although music is based on pitch relations rather than absolute pitch values, it is, paradoxically, these absolute pitch values that the brain is paying attention to throughout its different stages of processing.

A *scale* is just a subset of the theoretically infinite number of pitches, and every culture selects these based on historical tradition or somewhat arbitrarily. The specific pitches chosen are then anointed as being part of that musical system. These are the letters that you see in the figure above. The names "A," "B," "C," and so on are arbitrary labels that we associate with particular frequencies. In Western music—music of the European tradition—these pitches are the only "legal" pitches; most instruments are designed to play these pitches and not others. (Instruments like the trombone and cello are an exception, because they can slide between notes; trombonists, cellists, violinists, etc., spend a lot of time learning how to hear and produce the precise frequencies required to play each of the legal notes.) Sounds in between are considered mistakes ("out of tune") unless they're used for expressive intonation (intentionally playing something out of tune, briefly, to add emotional tension) or in passing from one legal tone to another.

Tuning refers to the precise relationship between the frequency of a tone being played and a standard, or between two or more tones being played together. Orchestral musicians "tuning up" before a performance are synchronizing their instruments (which naturally drift in their tuning as the wood, metal, strings, and other materials expand and contract with changes in temperature and humidity) to a standard frequency, or occasionally not to a standard but to each other. Expert musicians often alter the frequency of tones while they're playing for expressive purposes (except, of course, on fixed-pitch instruments such as keyboards and xylophones); sounding a note slightly lower or higher than its nominal value can impart emotion when done skillfully. Expert musicians playing together in ensembles will also alter the pitch of tones they play to bring them more in tune with the tones being played by the other musicians, should one or more musicians drift away from standard tuning during the performance.

The note names in Western music run from A to G, or, in an alternative system, as Do - re - mi - fa - sol - la - ti - do (the alternate system is used as lyrics to the Rodgers and Hammerstein song "Do-Re-Mi" from *The Sound of Music:* "Do, a deer, a female deer, Re, a drop of golden sun . . ."). As frequencies get higher, so do the letter names; B has a higher frequency than A (and hence a higher pitch) and C has a higher frequency than either A or B. After G, the note names start all over again at A. Notes with the same name have frequencies that are multiples of each other. One of the several notes we call A has a frequency of 55 Hz and all other notes called A have frequencies that are two, three, four, five (or a half) times this frequency.

Here is a fundamental quality of music. Note names repeat because of a perceptual phenomenon that corresponds to the doubling and halving of frequencies. When we double or halve a frequency, we end up with a note that sounds remarkably similar to the one we started out with. This relationship, a frequency ratio of 2:1 or 1:2, is called the octave. It is so important that, in spite of the large differences that exist between musical cultures—between Indian, Balinese, European, Middle Eastern, Chinese, and so on—every culture we know of has the octave as the basis for its music, even if it has little else in common with other musical traditions. This phenomenon leads to the notion of circularity in pitch perception, and is similar to circularity in colors. Although red and violet fall at opposite ends of the continuum of visible frequencies of electromagnetic energy, we see them as perceptually similar. The same is true in music, and music is often described as having two dimensions, one that accounts for tones going up in frequency (and sounding higher and higher) and another that accounts for the perceptual sense that we've come back home again each time we double a tone's frequency.

When men and women speak in unison, their voices are normally an octave apart, even if they try to speak the exact same pitches. Children generally speak an octave or two higher than adults. The first two notes of the Harold Arlen melody "Somewhere Over the Rainbow" (from the movie *The Wizard of Oz*) make an octave. In "Hot Fun in the Summertime" by Sly and the Family Stone, Sly and his backup singers are singing in octaves during the first line of the verse "End of the spring and here she comes back." As we increase frequencies by playing the successive notes on an instrument, there is a very strong perceptual sense that when we reach a doubling of frequency, we have come "home" again. The octave is so basic that even some animal species—monkeys and cats, for example—show octave equivalence, the ability to treat as similar, the way that humans do, tones separated by this amount.

An *interval* is the distance between two tones. The octave in Western music is subdivided into twelve (logarithmically) equally spaced tones.

The intervallic distance between A and B (or between "do" and "re") is called a whole step or a tone. (This latter term is confusing, since we call any musical sound a tone; I'll use the term *whole step* to avoid ambiguity). The smallest division in our Western scale system cuts a whole step perceptually in half: This is the semitone, which is one twelfth of an octave.

Intervals are the basis of melody, much more so than the actual pitches of notes; melody processing is relational, not absolute, meaning that we define a melody by its intervals, not the actual notes used to create them. Four semitones always create the interval known as a major third regardless of whether the first note is an A or a G# or any other note. Here is a table of the intervals as they're known in our (Western) musical system:

The table could continue on: Thirteen semitones is a minor ninth,

Distance in semitones	Interval name
0	unison
1	minor second
2	major second
3	minor third
4	major third
5	perfect fourth
6	augmented fourth, diminished fifth, or tritone
7	perfect fifth
8	minor sixth
9	major sixth
10	minor seventh
11	major seventh
12	octave

fourteen semitones is a major ninth, etc., but these names are typically used only in more advanced discussions. The intervals of the perfect fourth and perfect fifth are so called because they sound particularly pleasing to many people, and since the ancient Greeks, this particular feature of the scale is at the heart of all music. (There is no "imperfect fifth," this is just the name we give the interval.) Ignore the perfect fourth and fifth or use them in every phrase, they have been the backbone of music for at least five thousand years.

Although the areas of the brain that respond to individual pitches have been mapped, we have not yet been able to find the neurological basis for the encoding of pitch relations; we know which part of the cortex is involved in listening to the notes C and E, for example, and for F and A, but we do not know how or why both intervals are perceived as a major third, or the neural circuits that create this perceptual equivalency. These relations must be extracted by computational processes in the brain that remain poorly understood.

If there are twelve named notes within an octave, why are there only seven letters (or do-re-mi syllables)? After centuries of being forced to eat in the servants' quarters and to use the back entrance of the castle, this may just be an invention by musicians to make nonmusicians feel inadequate. The additional five notes have compound names, such as E♭ pronounced "E-flat") and F# (pronounced "F-sharp"). There is no reason for the system to be so complicated, but it is what we're stuck with.

The system is a bit clearer looking at the piano keyboard. A piano has white keys and black keys spaced out in an uneven arrangement—sometimes two white keys are adjacent, sometimes they have a black key between them. Whether the keys are white or black, the perceptual distance from one adjacent key to the next always makes a semitone, and a distance of two keys is always a whole step. This applies to many Western instruments; the distance between one fret on a guitar and the next is also a semitone, and pressing or lifting adjacent keys on woodwind instruments (such as the clarinet or oboe) typically changes the pitch by a semitone.

The white keys are named A, B, C, D, E, F, and G. The notes be-tween—the black keys—are the ones with compound names. The note between A and B is called either A-sharp or B-flat, and in all but formal music theoretic discussions, the two terms are interchangeable. (In fact, this note could also be referred to as C double-flat, and similarly, A could be called G double-sharp, but this is an even more theoretical usage.) Sharp means high, and flat means low. B-flat is the note one semitone lower than B; A-sharp is the note one semitone higher than A. In the par-allel do-re-mi system, unique syllables mark these other tones: di and ra indicate the tone between do and re, for example.

The notes with compound names are not in any way second-class mu-sical citizens. They are just as important, and in some songs and some scales they are used exclusively. For example, the main accompaniment to "Superstition" by Stevie Wonder is played on only the black keys of the keyboard. The twelve tones taken together, plus their repeating cousins one or more octaves apart, are the basic building blocks for melody, for all the songs in our culture. Every song you know, from "Deck the Halls" to "Hotel California," from "Ba Ba Black Sheep" to the theme from *Sex and the City*, is made up from a combination of these twelve tones and their octaves.

To add to the confusion, musicians also use the terms *sharp* and *flat* to indicate if someone is playing out of tune; if the musician plays the tone a bit too high (but not so high as to make the next note in the scale) we say that the tone being played is sharp, and if the musician plays the tone too low we say that the tone is flat. Of course, a musician can be only slightly off and nobody would notice. But when the musician is off by a relatively large amount—say one quarter to one half the distance be-tween the note she was trying to play and the next one—most of us can usually detect this and it sounds off. This is especially apparent when there is more than one instrument playing, and the out-of-tune tone we are hearing clashes with in-tune tones being played simultaneously by other musicians.

The names of pitches are associated with particular frequency values. Our current system is called A440 because the note we call A that is in the middle of the piano keyboard has been fixed to have a frequency of 440 Hz. This is entirely arbitrary. We could fix A at any frequency, such as 439, 444, 424, or 314.159; different standards were used in the time of Mozart than today. Some people claim that the precise frequencies affect the overall sound of a musical piece and the sound of instruments. Led Zeppelin often tuned their instruments away from the modern A440 standard to give their music an uncommon sound, and perhaps to link it with the European children's folk songs that inspired many of their compositions. Many purists insist on hearing baroque music on period instruments, both because the instruments have a different sound and because they are designed to play the music in its original tuning standard, something that purists deem important.

We can fix pitches anywhere we want because what defines music is a set of pitch relations. The specific frequencies for notes may be arbitrary, but the distance from one frequency to the next—and hence from one note to the next in our musical system—isn't at all arbitrary. Each note in our musical system is equally spaced to our ears (but not necessarily to the ears of other species). Although there is not an equal change in cycles per second (Hz) as we climb from one note to the next, the distance between each note and the next sounds equal. How can this be? The frequency of each note in our system is approximately 6 percent more than the one before it. Our auditory system is sensitive both to relative changes and to proportional changes in sound. Thus, each increase in frequency of 6 percent gives us the impression that we have increased pitch by the same amount as we did last time.

The idea of proportional change is intuitive if you think about weights. If you're at a gym and you want to increase your weight lifting of the barbells from 5 pounds to 50 pounds, adding 5 pounds each week is not going to change the amount of weight you're lifting in an equal way. After a week of lifting 5 pounds, when you move to 10 you are doubling the weight; the next week when you move to 15 you are adding 1.5 times as much weight as you had before. An equal spacing—to give your muscles a similar increase of weight each week—would be to add a constant percentage of the previous week's weight each time you increase.

For example, you might decide to add 50 percent each week, and so you would then go from 5 pounds to 7.5, then to 11.25, then to 16.83, and so on. The auditory system works the same way, and that is why our scale is based on a proportion: Every tone is 6 percent higher than the previous one, and when we increase each step by 6 percent twelve times, we end up having doubled our original frequency (the actual proportion is the twelfth root of two = 1.059463 . . .).

The twelve notes in our musical system are called the chromatic scale. Any scale is simply a set of musical pitches that have been chosen to be distinguishable from each other and to be used as the basis for constructing melodies.

In Western music we rarely use all the notes of chromatic scale in composition; instead, we use a subset of seven (or less often, five) of those twelve tones. Each of these subsets is itself a scale, and the type of scale we use has a large impact on the overall sound of a melody, and its emotional qualities. The most common subset of seven tones used in Western music is called the major scale, or Ionian mode (reflecting its ancient Greek origins). Like all scales, it can start on any of the twelve notes, and what defines the major scale is the specific pattern or distance relationship between each note and its successive note. In any major scale, the pattern of intervals—pitch distances between successive keys—is: whole step, whole step, half step, whole step, whole step, whole step, half step.

Starting on C, the major scale notes are C - D - E - F - G - A - B - C, all white notes on the piano keyboard. All other major scales require one or more black notes to maintain the required whole step/half step pattern. The starting pitch is also called the root of the scale.

The particular placement of the two half steps in the sequence of the major is crucial; it is not only what defines the major scale and distinguishes it from other scales, but it is an important ingredient in musical expectations. Experiments have shown that young children, as well as adults, are better able to learn and memorize melodies that are drawn from scales that contain unequal distances such as this. The presence of the two half steps, and their particular positions, orient the experienced, acculturated listener to where we are in the scale. We are all experts in knowing, when we hear a B in the key of C—that is, when the tones are being drawn primary from the C major scale—that it is the seventh note (or "degree") of that scale, and that it is only a half step below the root, even though most of us can't name the notes, and may not even know what a root or a scale degree is. We have assimilated the structure of this and other scales through a lifetime of listening and passive (rather than theoretically driven) exposure to the music. This knowledge is not innate, but is gained through experience. By a similar token, we don't need to know anything about cosmology to have learned that the sun comes up every morning and goes down at night—we have learned this sequence of events through largely passive exposure.

Different patterns of whole steps and half steps give rise to alternative scales, the most common of which (in our culture) is the minor scale. There is one minor scale that, like the C major scale, uses only the white notes of the piano keyboard: the A minor scale. The pitches for that scale are A - B - C - D - E - F - G - A. (Because it uses the same set of pitches, but in a different order, A minor is said to be the "relative minor of the C major scale.") The pattern of whole steps and half steps is different from that of the major scale: whole–half–whole–whole–half–whole–whole. Notice that the placement of the half steps is very different than in the major scale; in the major scale, there is a half step just before the root that "leads" to the root, and another half step just before the fourth scale degree. In the minor scale, the half steps are before the third scale degree and before the sixth. There is still a momentum when we're in this scale to return to the root, but the chords that create this momentum have a clearly different sound and emotional trajectory.

Now you might well ask: If these two scales use exactly the same set of pitches, how do I know which one I'm in? If a musician is playing the white keys, how do I know if he is playing the A minor scale or the C major scale? The answer is that—entirely without our conscious awareness—our brains are keeping track of how many times particular notes are sounded, where they appear in terms of strong versus weak beats, and how long they last. A computational process in the brain makes an inference about the key we're in based on these properties. This is another example of something that most of us can do even without musical training, and without what psychologists call declarative knowledge—the ability to talk about it; but in spite of our lack of formal musical education, we know what the composer intended to establish as the tonal center, or key, of the piece, and we recognize when he brings us back home to the tonic, or when he fails to do so. The simplest way to establish a key, then, is to play the root of the key many times, play it loud, and play it long. And even if a composer thinks he is writing in C major, if he has the musicians play the note A over and over again, play it loud and play it long; if the composer starts the piece on an A and ends the piece on an A, and moreover, if he avoids the use of C, the audience, musicians, and music theorists are most probably going to decide that the piece is in A minor, even if this was not his intent. In musical keys as in speeding tickets, it is the observed action, not the intention, that counts.

For reasons that are largely cultural, we tend to associate major scales with happy or triumphant emotions, and minor scales with sad or defeated emotions. Some studies have suggested that the associations might be innate, but the fact that these are not culturally universal indicates that, at the very least, any innate tendency can be overcome by exposure to specific cultural associations. Western music theory recognizes three minor scales and each has a slightly different flavor. Blues music generally uses a five note (pentatonic) scale that is a subset of the minor scale, and Chinese music uses a different pentatonic scale. When Tchaikovsky wants us to think of Arab or Chinese culture in the *Nutcracker* ballet, he chooses scales that are typical to their music, and within just a few notes we are transported to the Orient. When Billie Holiday wants to make a standard tune bluesy, she invokes the blues scale and sings notes from a scale that we are not accustomed to hearing in standard classical music.

Composers know these associations and use them intentionally. Our brains know them, too, through a lifetime of exposure to musical idioms, patterns, scales, lyrics, and the associations between them. Each time we hear a musical pattern that is new to our ears, our brains try to make an association through whatever visual, auditory and other sensory cues accompany it; we try to contextualize the new sounds, and eventually, we create these memory links between a particular set of notes and a particular place, time, or set of events. No one who has seen Hitchcock's *Psycho* can hear Bernard Hermann's screeching violins without thinking of the shower scene; anyone who has ever seen a Warner Bros. "Merrie Melody" cartoon will think of a character sneakily climbing stairs whenever they hear plucked violins playing an ascending major scale. The associations are so powerful—and the scales distinguishable enough—that only a few notes are needed: The first three notes of David Bowie's "China Girl" or Mussorgsky's "Great Gate of Kiev" (from *Pictures at an Exhibition*) instantly convey a rich and foreign (to us) musical context.

Nearly all this variation in context and sound comes from different ways of dividing up the octave and, in virtually every case we know of, dividing it up into no more than twelve tones. Although it has been claimed that Indian and Arab-Persian music use "microtuning"—scales with intervals much smaller than a semitone—close analysis reveals that their scales also rely on twelve or fewer tones and the others are simply expressive variations, glissandos (continuous glides from one tone to another), and momentary passing tones, similar to the American blues tradition of sliding into a note for emotional purposes.

In any scale, a hierarchy of importance exists among scale tones; some are more stable, structurally significant, or final sounding than others, causing us to feel varying amounts of tension and resolution. In the major scale, the most stable tone is the first degree, also called the tonic. In other words, all other tones in the scale seem to point toward the tonic, but they point with varying momentum. The tone that points most strongly to the tonic is the seventh scale degree, B in a C major scale. The tone that points least strongly to the tonic is the fifth scale degree, G in the C major scale, and it points least strongly because it is perceived as relatively stable; this is just another way of saying that we don't feel uneasy—unresolved—if a song ends on the fifth scale degree. Music theory specifies this tonal hierarchy. Carol Krumhansl and her colleagues performed a series of studies establishing that ordinary listeners have incorporated the principles of this hierarchy in their brains, through passive exposure to music and cultural norms. By asking people to rate how well different tones seemed to fit with a scale she would play them, she recovered from their subjective judgments the theoretical hierarchy.

A chord is simply a group of three or more notes played at the same time. They are generally drawn from one of the commonly used scales, and the three notes are chosen so that they convey information about the scale they were taken from. A typical chord is built by playing the first, third, and fifth notes of a scale together. Because the sequence of whole steps and half steps is different for minor and major scales, the interval sizes are different for chords taken in this way from the two different scales. If we build a chord starting on C and use the tones from the C major scale, we use C, E, and G. If instead we use the C minor scale, the first, third, and fifth notes are C, E-flat, and G. This difference in the third degree, between E and E-flat, turns the chord itself from a major chord into a minor chord. All of us, even without musical training, can tell the difference between these two even if we don't have the terminology to name them; we hear the major chord as sounding happy and the minor chord as sounding sad, or reflective, or even exotic. The most basic rock and country music songs use only major chords: "Johnny B. Goode," "Blowin' in the Wind," "Honky Tonk Women," and "Mammas Don't Let Your Babies Grow Up to Be Cowboys," for example.

Minor chords add complexity; in "Light My Fire" by the Doors, the verses are played in minor chords ("You know that it would be untrue . . .") and then the chorus is played in major chords ("Come on baby, light my fire"). In "Jolene," Dolly Parton mixes minor and major chords to give a melancholy sound. Pink Floyd's "Sheep" (from the album *Animals*) uses only minor chords.

Like single notes in the scale, chords also fall along a hierarchy of stability, depending on context. Certain chord progressions are part of every musical tradition, and even by the age of five, most children have internalized rules about what chord progressions are legal, or typical of their culture's music; they can readily detect deviations from the standard sequences just as easily as we can detect when an English sentence is malformed, such as this one: "The pizza was too hot to sleep." For brains to accomplish this, networks of neurons must form abstract representations of musical structure, and musical rules, something that they do automatically and without our conscious awareness. Our brains are maximally receptive—almost spongelike—when we're young, hungrily soaking up any and all sounds they can and incorporating them into the very structure of our neural wiring. As we age, these neural circuits are somewhat less pliable, and so it becomes more difficult to incorporate, at a deep neural level, new musical systems, or even new linguistic systems.

Now the story about pitch becomes a bit more complicated, and it's all the fault of physics. But this complication gives rise to the rich spectrum of sounds we hear in different instruments. All natural objects in the world have several modes of vibration. A piano string actually vibrates at several different rates at once. The same thing is true of bells that we hit with a hammer, drums that we hit with our hands, or flutes that we blow air into: The air molecules vibrate at several rates simultaneously, not just a single rate.

An analogy is the several types of motion of the earth that are simultaneously occurring. We know that the earth spins on its axis once every twenty-four hours, that it travels around the sun once every 365.25 days, and that the entire solar system is spinning along with the Milky Way galaxy. Several types of motion, all occurring at once. Another analogy is the many kinds of vibration that we often feel when riding a train. Imagine that you're sitting on a train in an outdoor station, with the engine off. It's windy, and you feel the car rock back and forth just a little bit. It does so with a regularity that you can time with your handy stopwatch, and you feel the train moving back and forth about twice a second. Next, the engineer starts the engine, and you feel a different kind of vibration through your seat (due to the oscillations of the motor—pistons and crankshafts turning around at a certain speed). When the train starts moving, you experience a third sensation, the bump the wheels make every time they go over a track joint. Altogether, you will feel several different kinds of vibrations, all of them likely to be at different rates, or frequencies. When the train is moving, you are no doubt aware that there is vibration. But it is very difficult, if not impossible, for you to determine how many vibrations there are and what their rates are. Using specialized measuring instruments, however, one might be able to figure this out.

When a sound is generated on a piano, flute, or any other instrument—including percussion instruments like drums and cowbells—it produces many modes of vibration occurring simultaneously. When you listen to a single note played on an instrument, you're actually hearing many, many pitches at once, not a single pitch. Most of us are not aware of this consciously, although some people can train themselves to hear this. The one with the slowest vibration rate—the one lowest in pitch— is referred to as the fundamental frequency, and the others are collectively called overtones.

To recap, it is a property of objects in the world that they generally vibrate at several different frequencies at once. Surprisingly, these other frequencies are often mathematically related to each other in a very simple way: as integer multiples of one another. So if you pluck a string and its slowest vibration frequency is one hundred times per second, the other vibration frequencies will be 2 x 100 (200 Hz), 3 x 100 Hz (300 Hz), etc. If you blow into a flute or recorder and cause vibrations at 310 Hz, additional vibrations will be occurring at twice, three times, four times, etc., this rate: 620 Hz, 930 Hz, 1240 Hz, etc. When an instrument creates energy at frequencies that are integer multiples such as this, we say that the sound is harmonic, and we refer to the pattern of energy at different frequencies as the overtone series. There is evidence that the brain responds to such harmonic sounds with synchronous neural firings—the neurons in auditory cortex responding to each of the components of the sound synchronize their firing rates with one another, creating a neural basis for the coherence of these sounds.

The brain is so attuned to the overtone series that if we encounter a sound that has all of the components except the fundamental, the brain fills it in for us in a phenomenon called *restoration of the missing fundamental*. A sound composed of energy at 100 Hz, 200 Hz, 300 Hz, 400

Hz, and 500 Hz is perceived as having a pitch of 100 Hz, its fundamental frequency. But if we artificially create a sound with energy at 200 Hz, 300 Hz, 400 Hz, and 500 Hz (leaving off the fundamental), we still perceive it as having a pitch of 100 Hz. We don't perceive it as having a pitch of 200 Hz, because our brain "knows" that a normal, harmonic sound with a pitch of 200 Hz would have an overtone series of 200 Hz, 400 Hz, 600 Hz, 800 Hz, etc. We can also fool the brain by playing sequences that deviate from the overtone series such as this: 100 Hz, 210 Hz, 302 Hz, 405 Hz, etc. In cases like these, the perceived pitch shifts away from 100 Hz in a compromise between what is presented and what a normal harmonic series would imply.

When I was in graduate school, my advisor, Mike Posner, told me about the work of a graduate student in biology, Petr Janata. Although he hadn't been raised in San Francisco like me, Petr had long bushy hair that he wore in a ponytail, played jazz and rock piano, and dressed in tie-dye: a true kindred spirit. Peter placed electrodes in the inferior colliculus of the barn owl, part of its auditory system. Then, he played the owls a version of Strauss's "The Blue Danube Waltz" made up of tones from which the fundamental frequency had been removed. Petr hypothesized that if the missing fundamental is restored at early levels of auditory processing, neurons in the owl's inferior colliculus should fire at the rate of the missing fundamental. This was exactly what he found. And because the electrodes put out a small electrical signal with each firing—and because the firing *rate* is the same as a *frequency* of firing—Petr sent the output of these electrodes to a small amplifier, and played back the sound of the owl's neurons through a loudspeaker. What he heard was astonishing; the melody of "The Blue Danube Waltz" sang clearly from the loudspeakers: ba da da da da, deet deet, deet deet. We were *hearing* the firing rates of the neurons and they were identical to the frequency of the missing fundamental. The overtone series had an instantiation not just in the early levels of auditory processing, but in a completely different species.

One could imagine an alien species that does not have ears, or that doesn't have the same internal experience of hearing that we do. But it would be difficult to imagine an advanced species that had no ability whatsoever to sense vibrating objects. Where there is atmosphere there are molecules that vibrate in response to movement. And knowing whether something is generating noise or moving toward us or away from us, even when we can't see it (because it is dark, our eyes aren't attending to it, or we're asleep) has a great survival value.

Because most physical objects cause molecules to vibrate in several modes at once, and because for many, many objects the modes bear simple integer relations to one another, the overtone series is a fact-of-the-world that we expect to find everywhere we look: in North America, in Fiji, on Mars, and on the planets orbiting Antares. Any organism that evolved in a world with vibrating objects is likely—given enough evolutionary time—to have evolved a processing unit in the brain that incorporated these regularities of its world. Because pitch is a fundamental cue to an object's identity, we would expect to find tonotopic mappings as we do in human auditory cortex, and synchronous neural firings for tones that bear octave and other harmonic relations to one another; this would help the brain (alien or terrestrial) to figure out that all these tones probably originated from the same object.

The overtones are often referred to by numbers: The first overtone is the first vibration frequency above the fundamental, the second overtone is the second vibration frequency above the fundamental, etc. Because physicists like to make the world confusing for the rest of us, there is a parallel system of terminology called harmonics, and I think it was designed to make undergraduates go crazy. In the lingo of harmonics, the first harmonic is the fundamental frequency, the second harmonic is equal to the first overtone, and so on. Not all instruments vibrate in modes that are so neatly defined. Sometimes, as with the piano (because it is a percussive instrument), the overtones can be close, but not exact, multiples of the fundamental frequency, and this contributes to their characteristic sound. Percussion instruments, chimes, and other objects—depending on composition and shape—often have overtones that are clearly not integer multiples of the fundamental, and these are called partials or inharmonic overtones. Generally, instruments with inharmonic overtones lack the clear sense of pitch that we associate with harmonic instruments, and the cortical basis for this may relate to a lack of synchronous neural firing. But they still do have a sense of pitch, and we hear this most clearly when we can play inharmonic notes in succession. Although you may not be able to hum along with the sound of a single note played on a woodblock or a chime, we can play a recognizable melody on a set of woodblocks or chimes because our brain focuses on the changes in the overtones from one to another. This is essentially what is happening when we hear people playing a song on their cheeks.

A flute, a violin, a trumpet, and a piano can all play the same tone— that is, you can write a note on a musical score and each instrument will play a tone with an identical fundamental frequency, and we will (tend to) hear an identical pitch. But these instruments all sound very different from one another.

This difference is timbre (pronounced TAM-ber), and it is the most important and ecologically relevant feature of auditory events. The timbre of a sound is the principal feature that distinguishes the growl of a lion from the purr of a cat, the crack of thunder from the crash of ocean waves, the voice of a friend from that of a bill collector one is trying to dodge. Timbral discrimination is so acute in humans that most of us can recognize hundreds of different voices. We can even tell whether someone close to us—our mother, our spouse—is happy or sad, healthy or coming down with a cold, based on the timbre of that voice.

Timbre is a consequence of the overtones. Different materials have different densities. A piece of metal will tend to sink to the bottom of a pond; an identically sized and shaped piece of wood will float. Partly due to density, and partly due to size and shape, different objects also make different noises when you strike them with your hand, or gently tap them with a hammer. Imagine the sound that you'd hear if you tap a hammer (gently, please!) against a guitar—a hollow, wooden *plunk* sound. Or if you tap a piece of metal, like a saxophone—a tinny *plink*. When you tap these objects, the energy from the hammer causes the molecules within them to vibrate, to dance at several different frequencies, frequencies determined by the material the object is made out of, its size, and its shape. If the object is vibrating at, say, 100 Hz, 200 Hz, 300 Hz, 400 Hz, etc., the intensity of vibration doesn't have to be the same for each of these harmonics, and in fact, typically, it is not.

When you hear a saxophone playing a tone with a fundamental frequency of 220 Hz, you are actually hearing many tones, not just one. The other tones you hear are integer multiples of the fundamental: 440, 660, 880, 1200, 1420, 1640, etc. These different tones—the overtones—have different intensities, and so we hear them as having different loudnesses. The particular pattern of loudnesses for these tones is distinctive of the saxophone, and they are what give rise to its unique tonal color, its unique sound—its timbre. A violin playing the same written note (220 Hz) will have overtones at the same frequencies, but the pattern of how loud each one is with respect to the others will be different. Indeed, for each instrument, there exists a unique pattern of overtones. For one instrument, the second overtone might be louder than in another, while the fifth overtone might be softer. Virtually all of the tonal variation we hear—the quality that gives a trumpet its trumpetiness and that gives a piano its pianoness—comes from the unique way in which the loudnesses of the overtones are distributed.

Each instrument has its own overtone profile, which is like a fingerprint. It is a complicated pattern that we can use to identify the instrument. Clarinets, for example, are characterized by having relatively high amounts of energy in the odd harmonics—three times, five times, and seven times the multiples of the fundamental frequency, etc. (This is a consequence of their being a tube that is closed at one end and open at the other.) Trumpets are characterized by having relatively even amounts of energy in both the odd and the even harmonics (like the clarinet, the trumpet is also closed at one end and open at the other, but the mouthpiece and bell are designed to smooth out the harmonic series). A violin that is bowed in the center will yield mostly odd harmonics and accordingly can sound similar to a clarinet. But bowing one third of the way down the instrument emphasizes the third harmonic and its multiples: the sixth, the ninth, the twelfth, etc.

All trumpets have a timbral fingerprint, and it is readily distinguishable from the timbral fingerprint for a violin, piano, or even the human voice. To the trained ear, and to most musicians, there even exist differences among trumpets—all trumpets don't sound alike, nor do all pianos or all accordions. (Well, to me all accordions sound alike, and the sweetest, most enjoyable sound I can imagine is the sound they would make burning in a giant bonfire.) What distinguishes one particular piano from another is that their overtone profiles will differ slightly from each other, but not, of course, as much as they will differ from the profile for a harpsichord, organ, or tuba. Master musicians can hear the difference between a Stradivarius violin and a Guarneri within one or two notes. I can hear the difference between my 1956 Martin 000-18 acoustic guitar, my 1973 Martin D-18, and my 1996 Collings D2H very clearly; they sound like different instruments, even though they are all acoustic guitars; I would never confuse one with another. That is timbre.

Natural instruments—that is, acoustic instruments made out of real-world materials such as metal and wood—tend to produce energy at several frequencies at once because of the way the internal structure of their molecules vibrates. Suppose that I invent an instrument that, unlike any natural instruments we know of, produces energy at one, and only one, frequency. Let's call this hypothetical instrument a generator (because it can generate tones of specific frequencies). If I line up a bunch of generators, I could set each one of them to play a specific frequency corresponding to the overtone series for a particular instrument playing a particular tone. I could have a bank of these generators making sounds at 110, 220, 330, 440, 550, and 660 Hz, which would give the listener the impression of a 110 Hz tone played by a musical instrument. Furthermore, I could control the amplitude of each of my generators and make each of the tones play at a particular loudness, corresponding to the overtone profile of a natural musical instrument. If I did that, the resulting bank of generators would approximate the sound of a clarinet, or flute, or any other instrument I was trying to emulate.

Additive synthesis such as the above approach achieves a synthetic version of a musical-instrument timbre by adding together elemental sonic components of the sound. Many pipe organs, such as those found in churches, have a feature that will let you play around with this. On most pipe organs you press a key (or a pedal), which sends a blast of air through a metal pipe. The organ is constructed of hundreds of pipes of different sizes, and each one produces a different pitch, corresponding to its size, when air is shot through it; you can think of them as mechanical flutes, in which the air is supplied by an electric motor rather than by a person blowing. The sound that we associate with a church organ—its particular timbre—is a function of there being energy at several different frequencies at once, just as with other instruments. Each pipe of the organ produces an overtone series, and when you press a key on the organ keyboard, a column of air is blasted through more than one pipe at a time, giving a very rich spectrum of sounds. These supplementary pipes, in addition to the one that vibrates at the fundamental frequency of the tone you're trying to play, either produce tones that are integer multiples of the fundamental frequency, or are closely related to it mathematically and harmonically.

The organ player typically has control over which of these supplementary pipes he wants to blow air through by pulling and pushing levers, or drawbars, that direct the flow of air. Knowing that clarinets have a lot of energy in the odd harmonics of the overtone series, a clever organ player could simulate the sound of a clarinet by manipulating drawbars in such a way as to re-create the overtone series of that instrument. A little bit of 220 Hz here, a dash of 330 Hz, a dollop of 440 Hz, a heaping helping of 550 Hz, and *voilà!*—you've cooked yourself up a reasonable facsimile of an instrument.

Starting in the late 1950s, scientists began experimenting with building such synthesis capabilities into smaller, more compact electronic devices, creating a family of new musical instruments known collectively as synthesizers. By the 1960s, synthesizers could be heard on records by the Beatles (on "Here Comes the Sun" and "Maxwell's Silver Hammer") and Walter/Wendy Carlos *(Switched-On Bach)*, followed by groups who sculpted their sound around the synthesizer, such as Pink Floyd and Emerson, Lake and Palmer.

Many of these synthesizers used additive synthesis as I've described it here, and later ones used more complex algorithms such as wave guide synthesis (invented by Julius Smith at Stanford) and FM synthesis (invented by John Chowning at Stanford). But merely copying the overtone profile, while it can create a sound reminiscent of the actual instrument, yields a rather pale copy. There is more to timbre than just the overtone series. Researchers still argue about what this "more" is, but it is generally accepted that, in addition to the overtone profile, timbre is defined by two other attributes that give rise to a perceptual difference from one instrument to another: attack and flux.

Stanford University sits on a bucolic stretch of land just south of San Francisco and east of the Pacific Ocean. Rolling hills covered with pastureland lie to the west, and the fertile Central Valley of California is just an hour or so to the east, home of a large proportion of the world's raisins, cotton, oranges, and almonds. To the south, near the town of Gilroy, are vast fields of garlic. Also to the south is Castroville, known as the "artichoke capitol of the world." (I once suggested to the Castroville Chamber of Commerce that they change *capitol* to *heart*. The response was not enthusiastic.)

Stanford has become something of a second home for computer scientists and engineers who love music. John Chowning, who was well known as an avant-garde composer, has had a professorship in the music department there since the 1970s, and was among a group of pioneering composers at the time who were using the computer to create, store, and reproduce sounds in their compositions. Chowning later became the founding director of the Center for Computer Research in Music and Acoustics at Stanford, known as CCRMA (pronounced CAR-ma; insiders joke that the first *c* is silent). Chowning is warm and friendly. When I was an undergraduate at Stanford, he would put his hand on my shoulder and ask what I was working on. You got the feeling talking to a student was for him an opportunity to learn something. In the early 1970s, while fiddling with the computer and with sine waves—the sorts of artificial sounds that are made by computers and used as the building blocks of additive synthesis—Chowning noticed that changing the frequency of these waves as they were playing created sounds that were musical. By controlling these parameters just so, he was able to simulate the sounds of a number of musical instruments. This new technique became known as frequency modulation synthesis, or FM synthesis, and became embedded first in the Yamaha DX9 and DX7 line of synthesizers, which revolutionized the music industry from the moment of their introduction in 1983. FM synthesis democratized music synthesis. Before FM, synthesizers were expensive, clunky, and hard to control. Creating new sounds took a great deal of time, experimentation, and know-how. But with FM, any musician could obtain a convincing instrumental sound at the touch of a button. Songwriters and composers who could not afford to hire a horn section or an orchestra could now play around with these textures and sounds. Composers and orchestrators could test out arrangements before taking the time of an entire orchestra to see what worked and what didn't. New Wave bands like the Cars and the Pretenders, as well as mainstream artists like Stevie Wonder, Hall and Oates, and Phil Collins, started to use FM synthesis widely in their recordings. A lot of what we think of as "the eighties sound" in popular music owes its distinctiveness to the particular sound of FM synthesis.

With the popularization of FM came a steady stream of royalty income that allowed Chowning to build up CCRMA, attracting graduate students and top-flight faculty members. Among the first of many famous electronic music/music-psychology celebrities to come to CCRMA were John R. Pierce and Max Mathews. Pierce had been the vice president of research at the Bell Telephone Laboratories in New Jersey, and supervised the team of engineers who built and patented the transistor—and it was Pierce who named the new device (TRANSfer resISTOR). In his distinguished career, he also is credited with inventing the traveling wave vacuum tube, and launching the first telecommunications satellite, Telstar. He was also a respected science fiction writer under the pseudonym J. J. Coupling. Pierce created a rare environment in any industry or research lab, one in which the scientists felt empowered to do their best and in which creativity was highly valued. At the time, the Bell Telephone Company/AT&T had a complete monopoly on telephone service in the U.S. and a large cash reserve. Their laboratory was something of a playground for the very best and brightest inventors, engineers, and scientists in America. In the Bell Labs "sandbox," Pierce allowed his people to be creative without worrying about the bottom line or the applicability of their ideas to commerce. Pierce understood that the only way true innovation can occur is when people don't have to censor themselves and can let their ideas run free. Although only a small proportion of those ideas may be practical, and a smaller proportion still would become products, those that did would be innovative, unique, and potentially very profitable. Out of this environment came a number of innovations including lasers, digital computers, and the Unix operating system.

I first met Pierce in 1990 when he was already eighty and was giving lectures on psychoacoustics at CCRMA. Several years later, after I had earned my Ph.D. and moved back to Stanford, we became friends and would go out to dinner every Wednesday night and discuss research. He once asked me to explain rock and roll music to him, something he had never paid any attention to and didn't understand. He knew about my previous career in the music business, and he asked if I could come over for dinner one night and play six songs that captured all that was important to know about rock and roll. Six songs to capture all of rock and roll? I wasn't sure I could come up with six songs to capture the Beatles, let alone all of rock and roll. The night before he called to tell me that he had heard Elvis Presley, so I didn't need to cover that.

Here's what I brought to dinner:

1) "Long Tall Sally," Little Richard
2) "Roll Over Beethoven," the Beatles
3) "All Along the Watchtower," Jimi Hendrix
4) "Wonderful Tonight," Eric Clapton
5) "Little Red Corvette," Prince
6) "Anarchy in the U.K.," the Sex Pistols

A couple of the choices combined great songwriters with different performers. All are great songs, but even now I'd like to make some adjustments. Pierce listened and kept asking who these people were, what instruments he was hearing, and how they came to sound the way they did. Mostly, he said that he liked the timbres of the music. The songs themselves and the rhythms didn't interest him that much, but he found the timbres to be remarkable—new, unfamiliar, and exciting. The fluid romanticism of Clapton's guitar solo in "Wonderful Tonight," combined with the soft, pillowy drums. The sheer power and density of the Sex Pistols' brick-wall-of-guitars-and-bass-and-drums. The sound of a distorted electric guitar wasn't all that was new to Pierce. The ways in which instruments were combined to create a unified whole—bass, drums, electric and acoustic guitars, and voice—that was something he had never heard before. Timbre was what defined rock for Pierce. And it was a revelation to both of us.

The pitches that we use in music—the scales—have remained essentially unchanged since the time of the Greeks, with the exception of the development—really a refinement—of the equal tempered scale during the time of Bach. Rock and roll may be the final step in a millennium-long musical revolution that gave perfect fourths and fifths a prominence in music that had historically been been given only to the octave. During this time, Western music was largely dominated by pitch. For the past two hundred years or so, timbre has become increasingly important. A standard component of music across all genres is to restate a melody using different instruments—from Beethoven's Fifth and Ravel's "Bolero" to the Beatles' "Michelle" and George Strait's "All My Ex's Live in Texas." New musical instruments have been invented so that composers might have a larger palette of timbral colors from which to draw. When a country or popular singer stops singing and another instrument takes up the melody—even without changing it in any way—we find pleasurable the repetition of the same melody with a different timbre.

The avant-garde composer Pierre Schaeffer (pronounced Sheh-FEHR, using your best imitation of a French accent) performed some crucial experiments in the 1950s that demonstrated an important attribute of timbre in his famous "cut bell" experiments. Schaeffer recorded a number of orchestral instruments on tape. Then, using a razor blade, he cut the beginnings off of these sounds. This very first part of a musical instrument sound is called the attack; this is the sound of the initial hit, strum, bowing, or blowing that causes the instrument to make sound.

The gesture our body makes in order to create sound from an instrument has an important influence on the sound the instrument makes. But most of that dies away after the first few seconds. Nearly all of the gestures we make to produce a sound are impulsive—they involve short, punctuated bursts of activity. In percussion instruments, the musician typically does not remain in contact with the instrument after this initial burst. In wind instruments and bowed instruments, on the other hand, the musician continues to be in contact with the instrument after the initial impulsive contact—the moment when the air burst first leaves her mouth or the bow first contacts the string; the continued blowing and bowing has a smooth, continuous, and less impulsive quality.

The introduction of energy to an instrument—the attack phase—usually creates energy at many different frequencies that are not related to one another by simple integer multiples. In other words, for the brief period after we strike, blow into, pluck, or otherwise cause an instrument to start making sound, the impact itself has a rather noisy quality that is not especially musical—more like the sound of a hammer hitting a piece of wood, say, than like a hammer hitting a bell or a piano string, or like the sound of wind rushing through a tube. Following the attack is a more stable phase in which the musical tone takes on the orderly pattern of overtone frequencies as the metal or wood (or other material) that the instrument is made out of starts to resonate. This middle part of a musical tone is referred to as the steady state—in most instances the overtone profile is relatively stable while the sound emanates from the instrument during this time.

After Schaeffer edited out the attack of orchestral instrument recordings, he played back the tape and found that it was nearly impossible for most people to identify the instrument that was playing. Without the attack, pianos and bells sounded remarkably unlike pianos and bells, and remarkably similar to one another. If you splice the attack of one instrument onto the steady state, or body, from another, you get varied results: In some cases, you hear an ambiguous hybrid instrument that sounds more like the instrument that the attack came from than the one the steady state came from. Michelle Castellengo and others have discovered that you can create entirely new instruments this way; for example, splicing a violin bow sound onto a flute tone creates a sound that strongly resembles a hurdy-gurdy street organ. These experiments showed the importance of the attack.

The third dimension of timbre—flux—refers to how the sound changes after it has started playing. A cymbal or gong has a lot of flux—its sound changes dramatically over the time course of its sound—while a trumpet has less flux—its tone is more stable as it evolves. Also, instruments don't sound the same across their range. That is, the timbre of an instrument sounds different when playing high and low notes. When Sting reaches up toward the top of his vocal range in "Roxanne" (by The Police), his straining, reedy voice conveys a type of emotion that he can't achieve in the lower parts of his register, such as we hear on the opening verse of "Every Breath You Take," a more deliberate, longing sound. The high part of Sting's register pleads with us urgently as his vocal cords strain, the low part suggests a dull aching that we feel has been going on for a long time, but has not yet reached the breaking point.

Timbre is more than the different sounds that instruments make. Composers use timbre as a compositional tool; they choose musical instruments—and combinations of musical instruments—to express particular emotions, and to convey a sense of atmosphere or mood. There is the almost comical timbre of the bassoon in Tchaikovsky's *Nutcracker Suite* as it opens the "Chinese Dance," and the sensuousness of Stan Getz's saxophone on "Here's That Rainy Day." Substitute a piano for the electric guitars in the Rolling Stones' "Satisfaction" and you'd have an entirely different animal. Ravel used timbre as a compositional device in *Bolero*, repeating the main theme over and over again with different timbres; he did this after he suffered brain damage that impaired his ability to hear pitch. When we think of Jimi Hendrix, it is the timbre of his electric guitars and his voice that we are likely to recall the most vividly.

Composers such as Scriabin and Ravel talk about their works as sound paintings, in which the notes and melodies are the equivalent of shape and form, and the timbre is equivalent to the use of color and shading. Several popular songwriters—Stevie Wonder, Paul Simon, and Lindsey Buckingham—have described their compositions as sound paintings, with timbre playing a role equivalent to the one that color does in visual art, separating melodic shapes from one another. But one of the things that makes music different from painting is that it is dynamic, changing across time, and what moves the music forward are rhythm and meter. Rhythm and meter are the engine driving virtually all music, and it is likely that they were the very first elements used by our ancestors to make protomusics, a tradition we still hear today in tribal drumming, and in the rituals of various preindustrial cultures. While I believe timbre is now at the center of our appreciation of music, rhythm has held supreme power over listeners for much longer.

2. Foot Tapping

Discerning Rhythm, Loudness, and Harmony

I saw Sonny Rollins perform in Berkeley in 1977; he is one of the most melodic saxophone players of our time. Yet nearly thirty years later, while I can't remember any of the pitches that he played, I clearly remember some of the rhythms. At one point, Rollins improvised for three and a half minutes by playing the same one note over and over again with different rhythms and subtle changes in timing. All that power in one note! It wasn't his melodic innovation that got the crowd to their feet—it was rhythm. Virtually every culture and civilization considers movement to be an integral part of music making and listening. Rhythm is what we dance to, sway our bodies to, and tap our feet to. In so many jazz performances, the part that excites the audience most is the drum solo. It is no coincidence that making music requires the coordinated, rhythmic use of our bodies, and that energy be transmitted from body movements to a musical instrument. At a neural level, playing an instrument requires the orchestration of regions in our primitive, reptilian brain—the cerebellum and the brain stem—as well as higher cognitive systems such as the motor cortex (in the parietal lobe) and the planning regions of our frontal lobes, the most advanced region of the brain.

Rhythm, meter, and tempo are related concepts that are often confused with one another. Briefly, *rhythm* refers to the lengths of notes, *tempo* refers to the pace of a piece of music (the rate at which you would tap your foot to it), and *meter* refers to when you tap your foot hard versus light, and how these hard and light taps group together to form larger units.

One of the things we usually want to know when performing music is how long a note is to be played. The relationship between the length of one note and another is what we call rhythm, and it is a crucial part of what turns sounds into music. Among the most famous rhythms in our culture is the rhythm often called "shave-and-a-haircut, two bits," sometimes used as the "secret" knock on a door. An 1899 recording by Charles Hale, "At a Darktown Cakewalk," is the first documented use of this rhythm. Lyrics were later attached to the rhythm in a song by Jimmie Monaco and Joe McCarthy called "Bum-Diddle-De-Um-Bum, That's It!" in 1914. In 1939, the same musical phrase was used in the song "Shave and a Haircut— Shampoo" by Dan Shapiro, Lester Lee, and Milton Berle. How the word *shampoo* became *two-bits* is a mystery. Even Leonard Bernstein got into the act by scoring this rhythm in the song "Gee, Officer Krupke" from the musical *West Side Story*. In "shave-and-a-haircut" we hear a series of notes of two different lengths, long and short; the long notes are twice as long as the short ones: long-short-short-long-long (rest) long-long.

In the *William Tell* overture by Rossini (what many of us know as the theme from *The Lone Ranger*) we also hear a series of notes of two different lengths, long and short; again, the long notes are twice as long as the short ones: da-da-bump da-da-bump da-da-bump bump bump (here I've used the "da" syllable for short, and the "bump" syllable for long). "Mary Had a Little Lamb" uses short and long syllables, too, in this case six equal duration notes (Ma-ry had a lit-tle) followed by a long one (lamb) roughly twice as long as the short ones. The rhythmic ratio of 2:1, like the octave in pitch ratios, appears to be a musical universal. We see it in the theme from *The Mickey Mouse Club* (bump-ba bump-ba bump-ba bump-ba bump-ba bump-ba baaaaah) in which we have three levels of duration, each one twice as long as the other. We see it in The Police's "Every Breath You Take" (da-da-bump da-da baaaaah), in which there are again three levels:

Ev-ry breath you-oo taaake
1   1   2   2   4

(The 1 represents one unit of some arbitrary time just to illustrate that the words *breath* and *you* are twice as long as the syllables *Ev* and *ry*, and that the word *take* is four times as long as *Ev* or *ry* and twice as long as *breath* or *you*.)

Rhythms in most of the music we listen to are seldom so simple. In the same way that a particular arrangement of pitches—the scale—can evoke music of a different culture, style, or idiom, so can a particular arrangement of rhythms. Although most of us couldn't reproduce a complex Latin rhythm, we recognize as soon as we hear it that it is Latin, as opposed to Chinese, Arabic, Indian, or Russian. When we organize rhythms into strings of notes, of varying lengths and emphases, we develop meter and establish tempo.

Tempo refers to the pace of a musical piece—how quickly or slowly it goes by. If you tap your foot or snap your fingers in time to a piece of music, the tempo of the piece will be directly related to how fast or slow you are tapping. If a song is a living, breathing entity, you might think of the tempo as its gait—the rate at which it walks by—or its pulse—the rate at which the heart of the song is beating. The word *beat* indicates the basic unit of measurement in a musical piece; this is also called the *tactus*. Most often, this is the natural point at which you would tap your foot or clap your hands or snap your fingers. Sometimes, people tap at half or twice the beat, due to different neural processing mechanisms from one person to another as well as differences in musical background, experience, and interpretation of a piece. Even trained musicians can disagree on what the tapping rate should be. But they always agree on the underlying speed at which the piece is unfolding, also called tempo; the disagreements are simply about subdivisions or superdivisions of that underlying pace.

Paula Abdul's "Straight Up" and AC/DC's "Back in Black" have a tempo of 96, meaning that there are 96 beats per minute. If you dance to "Straight Up" or "Back in Black," it is likely that you will be putting a foot down 96 times per minute or perhaps 48, but not 58 or 69. In "Back in Black" you can hear the drummer playing a beat on his high-hat cymbal at the very beginning, steadily, deliberately, at precisely 96 beats per minute. Aerosmith's "Walk This Way" has a tempo of 112, Michael Jackson's "Billie Jean" has a tempo of 116, and the Eagles' "Hotel California" has a tempo of 75.

Two songs can have the same tempo but feel very different. In "Back in Black," the drummer plays his cymbal twice for every beat (eighth notes) and the bass player plays a simple, syncopated rhythm perfectly in time with the guitar. On "Straight Up" there is so much going on, it is difficult to describe it in words. The drums play a complex, irregular pattern with beats as fast as sixteenth notes, but not continuously—the "air" between drum hits imparts a sound typical of funk and hip-hop music. The bass plays a similarly complex and syncopated melodic line that sometimes coincides with and sometimes fills in the holes of the drum part. In the right speaker (or the right ear of headphones) we hear the only instrument that actually plays on the beat every beat—a Latin instrument called an *afuche* or *cabasa* that sounds like sandpaper or beans shaking inside a gourd. Putting the most important rhythm on a light, high-pitched instrument is an innovative rhythmic technique that turns upside down the normal rhythmic conventions. While all this is going on, synthesizers, guitar, and special percussion effects fly in and out of the song dramatically, emphasizing certain beats now and again to add excitement. Because it is hard to predict or memorize where many of these are, the song holds a certain appeal over many, many listenings.

Tempo is a major factor in conveying emotion. Songs with fast tempos tend to be regarded as happy, and songs with slow tempos as sad. Although this is an oversimplification, it holds in a remarkable range of circumstances, across many cultures, and across the lifespan of an individual. The average person seems to have a remarkable memory for tempo. In an experiment that Perry Cook and I published in 1996, we asked people to simply sing their favorite rock and popular songs from memory and we were interested to know how close they came to the actual tempo of the recorded versions of those songs. As a baseline, we considered how much variation in tempo the average person can detect; that turns out to be 4 percent. In other words, for a song with a tempo of 100 bpm, if the tempo varies between 96–100, most people, even some professional musicians, won't detect this small change (although most drummers would—their job requires that they be more sensitive to tempo than other musicians, because they are responsible for maintaining tempo when there is no conductor to do it for them). A majority of people in our study—nonmusicians—were able to sing songs within 4 percent of their nominal tempo.

The neural basis for this striking accuracy is probably in the cerebellum, which is believed to contain a system of timekeepers for our daily lives and to synchronize to the music we are hearing. This means that somehow, the cerebellum is able to remember the "settings" it uses for synchronizing to music as we hear it, and it can recall those settings when we want to sing a song from memory. It allows us to synchronize our singing with a memory of the last time we sang. The basal ganglia— what Gerald Edelman has called "the organs of succession"—are almost certainly involved, as well, in generating and shaping rhythm, tempo, and meter.

Meter refers to the way in which the pulses or beats are grouped together. Generally when we're tapping or clapping along with music, there are some beats that we feel more strongly than others. It feels as if the musicians play this beat louder and more heavily than the others. This louder, heavier beat is perceptually dominant, and other beats that follow it are perceptually weaker until another strong one comes in. Every musical system that we know of has patterns of strong and weak beats. The most common pattern in Western music is for the strong beats to occur once every 4 beats: STRONG-weak-weak-weak STRONG-weak-weak-weak. Usually the third beat in a four-beat pattern is somewhat stronger than the second and fourth: There is a hierarchy of beat strengths, with the first being the strongest, the third being next, followed by the second and fourth. Somewhat less often the strong beat occurs once in every three in what we call the "waltz" beat: STRONG-weak-weak STRONG-weak-weak. We usually count to these beats as well, in a way that emphasizes which one is the strong beat: ONE-two-three-four, ONE-two-three-four, or ONE-two-three, ONE-two-three.

Of course music would be boring if we only had these straight beats. We might leave one out to add tension. Think of "Twinkle, Twinkle Little Star," written by Mozart when he was six years old. The notes don't occur on every beat:

ONE-two-three-four
ONE-two-three-(rest)
ONE-two-three-four
ONE-two-three-(rest):

TWIN-kle twin-kle
LIT-tle star (rest)
HOW-I won-der
WHAT you are (rest).

A nursery rhyme written to this same tune, "Ba Ba Black Sheep" subdivides the beat. A simple ONE-two-three-four can be divided into smaller, more interesting parts:

BA ba black sheep
HAVE-you-any-wool?

Notice that each syllable in "have-you-any" goes by twice as fast as the syllables "ba ba black." The quarter notes have been divided in half, and we can count this as

ONE-two-three-four
ONE-and-two-and-three-(rest).

In "Jailhouse Rock," performed by Elvis Presley and written by two outstanding songwriters of the rock era, Jerry Leiber and Mike Stoller, the strong beat occurs on the first note Presley sings, and then every fourth note after that:

[Line 1:] WAR-den threw a party at the
[Line 2:] COUN-ty jail (rest) the
[Line 3:] PRIS-on band was there and they be-
[Line 4:] GAN to wail

In music with lyrics, the words don't always line up perfectly with the downbeats; in "Jailhouse Rock" part of the word *began* starts before a strong beat and finishes on that strong beat. Most nursery rhymes and simple folk songs, such as "Ba Ba Black Sheep" or "Frère Jacques," don't do this. This lyrical technique works especially well on "Jailhouse Rock" because in speech the accent is on the second syllable of *began*; spreading the word across lines like this gives the song additional momentum.

By convention in Western music, we have names for the note durations similar to the way we name musical intervals. A musical interval of a "perfect fifth" is a relative concept—it can start on any note, and then by definition, notes that are either seven semitones higher or seven semitones lower in pitch are considered a perfect fifth away from the starting note. The standard duration is called a whole note and it lasts four beats, regardless of how slow or how fast the music is moving—that is, irrespective of tempo. (At a tempo of sixty beats per minute—as in the *Funeral March*—each beat lasts one second, so a whole note would last four seconds.) A note with half the duration of a whole note is called, logically enough, a half note, and a note half as long as that is called a quarter note. For most music in the popular and folk tradition, the quarter note is the basic pulse—the four beats that I was referring to earlier are beats of a quarter note. We talk about such songs as being in 4/4 time: The numerator tells us that the song is organized into groups of four notes, and the denominator tells us that the basic note length is a quarter note. In notation and conversation, we refer to each of these groups of four notes as a measure or a bar. One measure of music in 4/4 time has four beats, where each beat is a quarter note. This does not imply that the only note duration in the measure is the quarter note. We can have notes of any duration, or rests—that is to say, no notes at all; the 4/4 indication is only meant to describe how we count the beats.

"Ba Ba Black Sheep" has four quarter notes in its first measure, and then eighth notes (half the duration of a quarter note) and a quarter note rest in the second measure. I've used the symbol | to indicate a quarter note, and ∟ to indicate an eighth note, and I've kept the spacing between syllables proportional to how much time is spent on them:

[measure 1:] ba   ba   black   sheep
            | | | |
[measure 2:] have you an- y   wool (rest)
            ∟   ∟   ∟ ∟ | |

You can see in the diagram that the eighth notes go by twice as fast as the quarter notes.

In "That'll Be the Day" by Buddy Holly, the song begins with a pickup note; the strong beat occurs on the next note and then every fourth note after that, just as in "Jailhouse Rock":

THAT'll be the day (rest) when
YOU say good-bye-yes;
THAT'll be the day (rest) when
YOU make me cry-hi; you
SAY you gonna leave (rest) you
KNOW it's a lie 'cause
THAT'll be the day-ay-
AY when I die.

Notice how, like Elvis, Holly cuts a word in two across lines (*day* in the last two lines). To most people, the tactus is four beats between downbeats of this song, and they would tap their feet four times from one downbeat to the next. Here, all caps indicate the downbeat as before, and **bold** indicates when you would tap your foot against the floor:

Well
THAT'*ll **be** the **day** **(rest)** when*
YOU *say good-****bye-yes****;*
THAT'*ll **be** the **day** **(rest)** when*
YOU *make* me ***cry-hi****; you*
SAY *you **gonna** leave **(rest)** you*
KNOW *it's **a lie** 'cause*
THAT'*ll **be** the **day-ay-***
AY *when* ***I die****.*

If you pay close attention to the song's lyrics and their relationship to the beat, you'll notice that a foot tap occurs in the middle of some of the beats. The first *say* on the second line actually begins before you put your foot down—your foot is probably up in the air when the word *say* starts, and you put your foot down in the middle of the word. The same thing happens with the word *yes* later in that line. Whenever a note anticipates a beat—that is, when a musician plays a note a bit earlier than the strict beat would call for—this is called syncopation. This is a very important concept that relates to expectation, and ultimately to the emotional impact of a song. The syncopation catches us by surprise, and adds excitement.

As with many songs, some people feel "That'll Be the Day" in half time; there's nothing wrong with this—it is another interpretation and a valid one—and they tap their feet twice in the same amount of time other people tap four times: once on the downbeat, and again two beats later.

The song actually begins with the word *Well* that occurs before a strong beat—this is called a pickup note. Holly uses two words, *Well, you*, as pickup notes to the verse, also, and then right after them we're in sync again with the downbeats:

[pick up]   Well, you
[line 1]   GAVE me all your lovin' and your
[line 2]   (REST) tur-tle dovin' (rest)
[line 3]   ALL your hugs and kisses and your
[line 4]   (REST) money too.

What Holly does here that is so clever is that he violates our expectations not just with anticipations, but by delaying words. Normally, there would be a word on every downbeat, as in children's nursery rhymes. But in lines two and four of the song, the downbeat comes and he's silent! This is another way that composers build excitement, by not giving us what we would normally expect.

When people clap their hands or snap their fingers with music, they sometimes quite naturally, and without training, keep time differently than they would do with their feet: They clap or snap not on the downbeat, but on the second beat and the fourth beat. This is the so-called backbeat that Chuck Berry sings about in his song "Rock and Roll Music."

John Lennon said that the essence of rock and roll songwriting for him was to "Just say what it is, simple English, make it rhyme, and put a backbeat on it." In "Rock and Roll Music" (which John sang with the Beatles), as on most rock songs, the backbeat is what the snare drum is playing: The snare drum plays only on the second and fourth beat of each measure, in opposition to the strong beat which is on one, and a secondary strong beat, on three. This backbeat is the typical rhythmic element of rock music, and Lennon used it a lot as in "Instant Karma" (*whack* below indicates where the snare drum is played in the song, on the backbeat):

Instant karma's gonna get you
(rest) *whack* (rest) *whack*
"Gonna knock you right on the head"
(rest) *whack* (rest) *whack*

    . . .

But we all *whack* shine *whack*
on *whack* (rest) *whack*
Like the moon *whack* and the stars *whack*
and the sun *whack* (rest) *whack*

In "We Will Rock You" by Queen, we hear what sounds like feet stamping on stadium bleachers twice in a row (boom-boom) and then hand-clapping (CLAP) in a repeating rhythm: boom-boom-CLAP, boom-boom-CLAP; the CLAP is the backbeat.

Imagine now the John Philip Sousa march, "The Stars and Stripes Forever." If you can hear it in your mind, you can tap your foot along with the mental rhythm. While the music goes "DAH-dah-ta DUM-dum dah DUM-dum dum-dum DUM," your foot will be tapping DOWN-up DOWN-up DOWN-up DOWN-up. In this song, it is natural to tap your foot for every two quarter notes. We say that this song is "in two," meaning that the natural grouping of rhythms is two quarter notes per beat.

Now imagine "My Favorite Things" (words and music by Richard Rodgers and Oscar Hammerstein). This song is in waltz time, or what is called 3/4 time. The beats seem to arrange themselves in groups of three, with a strong beat followed by two weak ones. "RAIN-drops-on ROSE-es and WHISK-ers-on KIT-tens (rest)." ONE-two-three ONE-two-three ONE-two-three ONE-two-three.

As with pitch, small-integer ratios of durations are the most common, and there is accumulating evidence that they are easier to process neurally. But, as Eric Clarke notes, small-integer ratios are almost never found in samples of real music. This indicates that there is a quantization process—equalizing durations—occurring during our neural processing of musical time. Our brains treat durations that are similar as being equal, rounding some up and some down in order to treat them as simple integer ratios such as 2:1, 3:1 and 4:1. Some musics use more complex ratios than these; Chopin and Beethoven use nominal ratios of 7:4 and and 5:4 in some of their piano works, in which seven or five notes are played with one hand while the other hand plays four. As with pitch, any ratio is theoretically possible, but there are limitations to what we can perceive and remember, and there are limitations based on style and convention.

The three most common meters in Western music are: 4/4, 2/4, and 3/4. Other rhythmic groupings exist, such as 5/4, 7/4, and 9/4. A somewhat common meter is 6/8, in which we count six beats to a measure, and each eighth note gets one beat. This is similar to 3/4 waltz time, the difference being that the composer intends for the musicians to "feel" the music in groups of six rather than groups of three, and for the underlying pulse to be the shorter-duration eighth note rather than a quarter note. This points to the hierarchy that exists in musical groupings. It is possible to count 6/8 as two groups of 3/8 (ONE-two-three ONE-two-three) or as one group of six (ONE-two-three-FOUR-five-six) with a secondary accent on the fourth beat, and to most listeners these are uninteresting subtleties that only concern a performer. But there may be brain differences. We know that there are neural circuits specifically related to detecting and tracking musical meter, and we know that the cerebellum is involved in setting an internal clock or timer that can synchronize with events that are out-there-in-the-world. No one has yet done the experiment to see if 6/8 and 3/4 have different neural representations, but because musicians truly treat them as different, there is a high probability that the brain does also. A fundamental principle of cognitive neuroscience is that the brain provides the biological basis for any behaviors or thoughts that we experience, and so at some level there must be neural differentiation wherever there is behavioral differentiation.

Of course, 4/4 and 2/4 time are easy to walk to, dance to, or march to because (since they are even numbers) you always end up with the same foot hitting the floor on a strong beat. Three-quarter is less natural to walk to; you'll never see a military outfit or infantry division marching to 3/4. Five-quarter time is used once in a while, the most famous examples being Lalo Shiffrin's theme from *Mission: Impossible*, and the Dave Brubeck song "Take Five." As you count the pulse and tap your foot to these songs, you'll see that the basic rhythms group into fives: ONE-two-three-four-five, ONE-two-three-four-five. There is a secondary strong beat in Brubeck's composition on the four: ONE-two-three-FOUR-five. In this case, many musicians think of 5/4 beats as consisting of alternating 3/4 and 2/4 beats. In "Mission: Impossible," there is no clear subdivision of the five. Tchaikovsky uses 5/4 time for the second movement of his Sixth Symphony. Pink Floyd used 7/4 for their song "Money," as did Peter Gabriel for "Salisbury Hill"; if you try to tap your foot or count along, you'll need to count seven between each strong beat.

I left discussion of loudness for almost-last, because there really isn't much to say about loudness in terms of definition that most people don't already know. One counterintuitive point is that loudness is, like pitch, an entirely psychological phenomenon, that is, loudness doesn't exist in the world, it only exists in the mind. And this is true for the same reason that pitch only exists in the mind. When you're adjusting the output of your stereo system, you're technically increasing the amplitude of the vibration of molecules, which in turn is interpreted as loudness by our brains. The point here is that it takes a brain to experience what we call "loudness." This may seem largely like a semantic distinction, but it is important to keep our terms straight. Several odd anomalies exist in the mental representation of amplitude, such as loudnesses not being additive the way that amplitudes are (loudness, like pitch, is logarithmic), or the phenomenon that the pitch of a sinusoidal tone varies as a function of its amplitude, or the finding that sounds can appear to be louder than they are when they have been electronically processed in certain ways—such as dynamic range compression—that are often done in heavy metal music.

Loudness is measured in decibels (named after Alexander Graham Bell and abbreviated dB) and it is a dimensionless unit like percent; it refers to a ratio of two sound levels. In this sense, it is similar to talking about musical intervals, but not to talking about note names. The scale is logarithmic, and doubling the intensity of a sound source results in a 3 dB increase in sound. The logarithmic scale is useful for discussing sound because of the ear's extraordinary sensitivity: The ratio between the loudest sound we can hear without causing permanent damage and the softest sound we can detect is a million to one, when measured as sound-pressure levels in the air; on the dB scale this is 120 dB. The range of loudnesses we can perceive is called the dynamic range. Sometimes critics talk about the dynamic range that is achieved on a high-quality music recording; if a record has a dynamic range of 90 dB, it means that the difference between the softest parts on the record and the loudest parts is 90 dB—considered high fidelity by most experts, and beyond the capability of most home audio systems.

Our ears compress sounds that are very loud in order to protect the delicate components of the middle and inner ear. Normally, as sounds get louder in the world, our perception of the loudness increases proportionately to them. But when sounds are really loud, a proportional increase in the signal transmitted by the eardrum would cause irreversible damage. The compression of the sound levels—of the dynamic range—means that large increases in sound level in the world create much smaller changes of level in our ears. The inner hair cells have a dynamic range of 50 decibels (dB) and yet we can hear over a 120 dB dynamic range. For every 4 dB increase in sound level, a 1 dB increase is transmitted to the inner hair cells. Most of us can detect when this compression is taking place; compressed sounds have a different quality.

Acousticians have developed a way to make it easy to talk about sound levels in the environment—because dBs express a ratio between two values, they chose a standard reference level (20 micropascals of sound pressure) which is approximately equal to the threshold of human hearing for most healthy people—the sound of a mosquito flying ten feet away. To avoid confusion, when decibels are being used to reflect this reference point of sound pressure level, we refer to them as dB (SPL). Here are some landmarks for sound levels, expressed in dB (SPL):

0 dB	Mosquito flying in a quiet room, ten feet away from your ears
20 dB	A recording studio or a very quiet executive office
35 dB	A typical quiet office with the door closed and computers off
50 dB	Typical conversation in a room

75 dB	Typical, comfortable music listening level in headphones
100–105 dB	Classical music or opera concert during loud passages; some portable music players go to 105 dB
110 dB	A jackhammer three feet away
120 dB	A jet engine heard on the runway from three hundred feet away; a typical rock concert
126–130 dB	Threshold of pain and damage; a rock concert by the Who (note that 126 dB is four times as loud as 120 dB)
180 dB	Space shuttle launch
250–275 dB	Center of a tornado; volcanic eruption

Conventional foam insert earplugs can block about 25 dB of sound, although they do not do so across the entire frequency range. Earplugs at a Who concert can minimize the risk of permanent damage by bringing down the levels that reach the ear close to 100–110 dB (SPL). The over-the-ear type of ear protector worn at rifle firing ranges and by airport landing personnel is often supplemented by in-the-ear plugs to afford maximum protection.

A lot of people like really loud music. Concertgoers talk about a special state of consciousness, a sense of thrills and excitement, when the music is really loud—over 115 dB. We don't yet know why this is so. Part of the reason may be related to the fact that loud music saturates the auditory system, causing neurons to fire at their maximum rate. When many, many neurons are maximally firing, this could cause an emergent property, a brain state qualitatively different from when they are firing at normal rates. Still, some people like loud music, and some people don't.

Loudness is one of the seven major elements of music along with pitch, rhythm, melody, harmony, tempo, and meter. Very tiny changes in loudness have a profound effect on the emotional communication of music. A pianist may play five notes at once and make one note only slightly louder than the others, causing it to take on an entirely different role in our overall perception of the musical passage. Loudness is also an important cue to rhythms, as we saw above, and to meter, because it is the loudness of notes that determines how they group rhythmically.

* * *

Now we have come full circle and return to the broad subject of pitch. Rhythm is a game of expectation. When we tap our feet we are predicting what is going to happen in the music next. We also play a game of expectations in music with pitch. Its rules are key and harmony. A musical key is the tonal context for a piece of music. Not all musics have a key. African drumming, for instance, doesn't, nor does the twelve-tone music of contemporary composers such as Schönberg. But virtually all of the music we listen to in Western culture—from commercial jingles on the radio to the most serious symphony by Bruckner, from the gospel music of Mahalia Jackson to the punk of the Sex Pistols—has a central set of pitches that it comes back to, a tonal center, the key. The key can change during the course of the song (called modulation), but by definition, the key is generally something that holds for a relatively long period of time during the course of the song, typically on the order of minutes.

If a melody is based on the C major scale, for example, we generally say that the melody is "in the key of C." This means that the melody has a momentum to return to the note C, and that even if it doesn't end on a C, the note C is what listeners are keeping in their minds as the dominant and focal note of the entire piece. The composer may temporarily use notes from outside the C major scale, but we recognize those as departures—something like a quick edit in a movie to a parallel scene or a flashback, in which we know that a return to the main plotline is imminent and inevitable. (For a more detailed look at music theory see Appendix 2.)

The attribute of pitch in music functions within a scale or a tonal/harmonic context. A note doesn't always sound the same to us every time we hear it: We hear it within the context of a melody and what has come before, and we hear it within the context of the harmony and chords that are accompanying it. We can think of it like flavor: Oregano tastes good with eggplant or tomato sauce, maybe less good with banana pudding. Cream takes on a different gustatory meaning when it is on top of strawberries from when it is in coffee or part of a creamy garlic salad dressing.

In "For No One" by the Beatles, the melody is sung on one note for two measures, but the chords accompanying that note change, giving it a different mood and a different sound. The song "One Note Samba" by Antonio Carlos Jobim actually contains many notes, but one note is featured throughout the song with changing chords accompanying it, and we hear a variety of different shades of musical meaning as this unfolds. In some chordal contexts, the note sounds bright and happy, in others, pensive. Another thing that most of us are expert in, even if we are nonmusicians, is recognizing familiar chord progressions, even in the absence of the well-known melody. Whenever the Eagles play this chord sequence in concert

B minor / F-sharp major / A major / E major / G major / D major / E minor / F-sharp major

they don't have to play more than three chords before thousands of nonmusician fans in the audience know that they are going to play "Hotel California." And even as they have changed the instrumentation over the years, from electric to acoustic guitars, from twelve-string to six-string guitars, people recognize those chords; we even recognize them when they're played by an orchestra coming out of cheap speakers in a Muzak version in the dentist's office.

Related to the topic of scales and major and minor is the topic of tonal consonance and dissonance. Some sounds strike us as unpleasant, although we don't always know why. Fingernails screeching on a chalkboard are a classic example, but this seems to be true only for humans; monkeys don't seem to mind (or at least in the one experiment that was done, they like that sound as much as they like rock music). In music, some people can't stand the sound of distorted electric guitars; others won't listen to anything else. At the harmonic level—that is, the level of the notes, rather than the timbres involved—some people find particular intervals or chords particularly unpleasant. Musicians refer to the pleasing-sounding chords and intervals as consonant and the unpleasing ones as dissonant. A great deal of research has focused on the problem of why we find consonant some intervals and not others, and there is currently no agreement about this. So far, we've been able to figure out that the brain stem and the dorsal cochlear nucleus—structures that are so primitive that all vertebrates have them—can distinguish between consonance and dissonance; this distinction happens before the higher level, human brain region—the cortex—gets involved.

Although the neural mechanisms underlying consonance and dissonance are debated, there is widespread agreement about some of the intervals that are deemed consonant. A unison interval—the same note played with itself—is deemed consonant, as is an octave. These create simple integer frequencies ratios of 1:1 and 2:1 respectively. (From an acoustics standpoint, half of the peaks in the waveform for octaves line up with each other perfectly, the other half fall exactly in between two peaks.) Interestingly, if we divide the octave precisely in half, the interval we end up with is called a tritone and most people find it the most disagreeable interval possible. Part of the reason for this may be related to the fact that the tritone does not come from a simple integer ratio, its ratio being 43:32. We can look at consonance from an integer ratio perspective. A ratio of 3:1 is a simple integer ratio, and that defines two octaves. A ratio of 3:2 is also a simple integer ratio, and that defines the interval of a perfect fifth. This is the distance between, for example, C and the G above it. The distance from that G to the C above it forms an interval of a perfect fourth, and its frequency ratio is 4:3.

The particular notes found in our major scale trace their roots back to the ancient Greeks and their notions of consonance. If we start with a note C and simply add the interval of a perfect fifth to it iteratively, we end up generating a set of frequencies that are very close to the current major scale: C - G - D - A - E - B - F-sharp - C-sharp - G-sharp - D-sharp - A-sharp - E-sharp (or F), and then back to C. This is known as the circle of fifths because after going through the cycle, we end up back at the note we started on. Interestingly, if we follow the overtone series, we can generate frequencies that are somewhat close to the major scale as well.

A single note cannot, by itself, be dissonant, but it can sound dissonant against the backdrop of certain chords, particularly when the chord implies a key that the single note is not part of. Two notes can sound dissonant together, both when played simultaneously or in sequence, if the sequence does not conform to the customs we have learned that go with our musical idioms. Chords can also sound dissonant, especially when they are drawn from outside the key that has been established. Bringing all these factors together is the task of the composer. Most of us are very discriminating listeners, and when the composer gets the balance just slightly wrong, our expectations have been betrayed more than we can stand, and we switch radio stations, pull off the earphones, or just walk out of the room.

I've reviewed the major elements that go into music: pitch, timbre, key, harmony, loudness, rhythm, meter, and tempo. Neuroscientists deconstruct sound into its components to study selectively which brain regions are involved in processing each of them, and musicologists discuss their individual contributions to the overall aesthetic experience of listening. But music—real music—succeeds or fails because of the relationship among these elements. Composers and musicians rarely treat these in total isolation; they know that changing a rhythm may also require changing pitch or loudness, or the chords that accompany that rhythm. One approach to studying the relationship between these elements traces its origins back to the late 1800s and the Gestalt psychologists.

In 1890, Christian von Ehrenfels was puzzled by something all of us take for granted and know how to do: melodic transposition. Transposition is simply singing or playing a song in a different key or with different pitches. When we sing "Happy Birthday" we just follow along with the first person who started singing, and in most cases, this person just starts on any note that she feels like. She might even have started on a pitch that is not a recognized note of the musical scale, falling between, say, C and C-sharp, and almost no one would notice or care. Sing "Happy Birthday" three times in a week and you might be singing three completely different sets of pitches. Each version of the song is called a transposition of the others.

The Gestalt psychologists—von Ehrenfels, Max Wertheimer, Wolfgang Köhler, Kurt Koffka, and others—were interested in the problem of configurations, that is, how it is that elements come together to form wholes, objects that are qualitatively different from the sum of their parts, and cannot be understood in terms of their parts. The word *Gestalt* has entered the English language to mean a unified whole form, applicable to both artistic and nonartistic objects. One can think of a suspension bridge as a Gestalt. The functions and utility of the bridge are not easily understood by looking at pieces of cable, girders, bolts, and steel beams; it is only when they come together in the form of a bridge that we can apprehend how a bridge is different from, say, a construction crane that might be made out of the same parts. Similarly, in painting, the relationship between elements is a critical aspect of the final artistic product. The classic example is a face—the *Mona Lisa* would not be what it is if the eyes, nose, and mouth were painted entirely as they are but were scattered across the canvas in a different arrangement.

The Gestaltists wondered how it is that a melody—composed of a set of specific pitches—could retain its identity, its recognizability, even when all of its pitches are changed. Here was a case for which they could not generate a satisfying theoretical explanation, the ultimate triumph of form over detail, of the whole over the parts. Play a melody using any set of pitches, and so long as the relation between those pitches is held constant, it is the same melody. Play it on different instruments and people still recognize it. Play it at half speed or double speed, or impose all of these transformations at the same time, and people still have no trouble recognizing it as the original song. The influential Gestalt school was formed to address this particular question. Although they never answered it, they did go on to contribute enormously to our understanding of how objects in the visual world are organized, through a set of rules that are taught in every introductory psychology class, the "Gestalt Principles of Grouping."

Albert Bregman, a cognitive psychologist at McGill University, has performed a number of experiments over the last thirty years to develop a similar understanding of grouping principles for sound. The music theorist Fred Lerdahl from Columbia University and the linguist Ray Jackendoff from Brandeis University (now at Tufts University) tackled the problem of describing a set of rules, similar to the rules of grammar in spoken language, that govern musical composition, and these include grouping principles for music. The neural basis for these principles has not been competely worked out, but through a series of clever behavioral experiments we have learned a great deal about the phenomenology of the principles.

In vision, grouping refers to the way in which elements in the visual world combine or stay separate from one another in our mental image of the world. Grouping is partly an automatic process, which means that much of it happens rapidly in our brains and without our conscious awareness. It has been described simply as the problem of "what goes with what" in our visual field. Hermann von Helmholtz, the nineteenth-century scientist who taught us much of what we now accept as the foundations of auditory science, described it as an unconscious process that involved inferencing, or logical deductions about what objects in the world are likely to go together based on a number of features or attributes of the objects.

If you're standing on a mountaintop overlooking a varied landscape, you might describe seeing two or three other mountains, a lake, a valley, a fertile plain, and a forest. Although the forest is composed of hundreds or thousands of trees, the trees form a perceptual group, distinct from other things we see, not necessarily because of our knowledge of forests, but because the trees share similar properties of shape, size, and color—at least when they stand in opposition to fertile plains, lakes, and mountains. But if you're in the center of a forest with a mixture of alder trees and pines, the smooth white bark of the alders will cause them to "pop out" as a separate group from the craggy dark-barked pines. If I put you in front of one tree and ask you what you see, you might start to focus on details of that tree: bark, branches, leaves (or needles), insects, and moss. When looking at a lawn, most of us don't typically see individual blades of grass, although we can if we focus our attention on them. Grouping is a hierarchical process and the way in which our brains form perceptual groups is a function of a great many factors. Some grouping factors are intrinsic to the objects themselves—shape, color, symmetry, contrast, and principles that address the continuity of lines and edges of the object. Other grouping factors are psychological, that is, mind based, such as what we're consciously trying to pay attention to, what memories we have of this or similar objects, and what our expectations are about how objects should go together.

Sounds group too. This is to say that while some group with one another, others segregate from each other. Most people can't isolate the sound of one of the violins in an orchestra from the others, or one of the trumpets from the others—they form a group. In fact, the entire orchestra can form a single perceptual group—called a stream in Bregman's terminology—depending on the context. If you're at an outdoor concert with several ensembles playing at once, the sounds of the orchestra in front of you will cohere into a single auditory entity, separate from the other orchestras behind you and off to the side. Through an act of volition (attention) you can then focus on just the violins of the orchestra in front of you, just as you can follow a conversation with the person next to you in a crowded room full of conversations.

One case of auditory grouping is the way that the many different sounds emanating from a single musical instrument cohere into a percept of a single instrument. We don't hear the individual harmonics of an oboe or of a trumpet, we hear an oboe or we hear a trumpet. This is all the more remarkable if you imagine an oboe and a trumpet playing at the same time. Our brains are capable of analyzing the dozens of different frequencies reaching our ears, and putting them together in just the right way. We don't have the impression of dozens of disembodied harmonics, nor do we hear just a single hybrid instrument. Rather, our brains construct for us separate mental images of an oboe and of a trumpet, and also of the sound of the two of them playing together—the basis for our appreciation of timbral combinations in music. This is what Pierce was talking about when he marveled at the timbres of rock music—the sounds that an electric bass and an electric guitar made when they were playing together—two instruments, perfectly distinguishable from one another, and yet simultaneously creating a new sonic combination that can be heard, discussed, and remembered.

Our auditory system exploits the harmonic series in grouping sounds together. Our brains coevolved in a world in which many of the sounds that our species encountered—over the tens of thousands of years of evolutionary history—shared certain acoustical properties with one another, including the harmonic series as we now understand it. Through this process of "unconscious inference" (as von Helmholtz called it), our brains assume that it is highly unlikely that several different sound sources are present, each producing a single component of the harmonic series. Rather, our brains use the "likelihood principle" that it must be a single object producing these harmonic components. All of us can make these inferences, even those of us who can't identify or name the instrument "oboe" as distinct, from, say, a clarinet or bassoon, or even a violin. But just as people who don't know the names of the notes in the scale can still tell when two different notes are being played as opposed to the same notes, nearly all of us—even lacking a knowledge of the names of musical instruments—can tell when there are two different instruments playing. The way in which we use the harmonic series to group sounds goes a long way toward explaining why we hear a trumpet rather the individual overtones that impinge on our ears—they group together like blades of grass that give us the impression of "lawn." It also explains how we can distinguish a trumpet from an oboe when they're each playing different notes—different fundamental frequencies give rise to a different set of overtones, and our brains are able to effortlessly figure out what goes with what, in a computational process that resembles what a computer might do. But it doesn't explain how we might be able to distinguish a trumpet from an oboe when they're playing the same note, because then the overtones are very nearly the same in frequency (although with different amplitudes characteristic of the instrument). For that, the auditory system relies on a principle of simultaneous onsets. Sounds that begin together—at the same instant in time—are perceived as going together, in the grouping sense. And it has been known since the time Wilhelm Wundt set up the first psychological laboratory in the 1870s that our auditory system is exquisitely sensitive to what constitutes simultaneous in this sense, being able to detect differences in onset times as short as a few milliseconds.

So when a trumpet and an oboe are playing the same note at the same time, our auditory system is able to figure out that two different instruments are playing because the full sound spectrum—the overtone series—for one instrument begins perhaps a few thousandths of a second before the sound spectrum for the other. This is what is meant by a grouping process that not only integrates sounds into a single object, but segregates them into different objects.

This principle of simultaneous onsets can be thought of more generally as a principle of temporal positioning. We group all the sounds that the orchestra is making now as opposed to those it will make tomorrow night. Time is a factor in auditory grouping. Timbre is another, and this is what makes it so difficult to distinguish one violin from several that are all playing at once, although expert musicians and conductors can train themselves to do this. Spatial location is a grouping principle, as our ears tend to group together sounds that come from the same relative position in space. We are not very sensitive to location in the up-down plane, but we are very sensitive to position in the left-right plane and somewhat sensitive to distance in the forward-back plane. Our auditory system assumes that sounds coming from a distinct location in space are probably part of the same object-in-the-world. This is one of the explanations for why we can follow a conversation in a crowded room relatively easily— our brains are using the cues of spatial location of the person we're conversing with to filter out other conversations. It also helps that the person we're speaking to has a unique timbre—the sound of his voice— that works as an additional grouping cue.

Amplitude also affects grouping. Sounds of a similar loudness group together, which is how we are able to follow the different melodies in Mozart's divertimenti for woodwinds. The timbres are all very similar, but some instruments are playing louder than others, creating different streams in our brains. It is as though a filter or sieve takes the sound of the woodwind ensemble and separates it out into different parts depending on what part of the loudness scale they are playing in.

Frequency, or pitch, is a strong and fundamental consideration in grouping. If you've ever heard a Bach flute partita, there are typically moments when some flute notes seem to "pop out" and separate themselves from one another, particularly when the flautist is playing a rapid passage—the auditory equivalent of a "Where's Waldo?" picture. Bach knew about the ability of large frequency differences to segregate sounds from one another—to block or inhibit grouping—and he wrote parts that included large leaps in pitch of a perfect fifth or more. The high notes, alternating with a succession of lower-pitched notes, create a separate stream and give the listener the illusion of two flutes playing when there is only one. We hear the same thing in many of the violin sonatas by Locatelli. Yodelers can accomplish the same effect with their voices, by combining pitch and timbral cues; when a male yodeler jumps into his falsetto register, he is creating both a distinct timbre and, typically, a large jump in pitch, causing the higher notes to again separate out into a distinct, perceptual stream, giving the illusion of two people singing interleaved parts.

We now know that the neurobiological subsystems for the different attributes of sound that I've described separate early on, at low levels of the brain. This suggests that grouping is carried out by general mechanisms working somewhat independently of one another. But it is also clear that the attributes work with or against each other when they combine in particular ways, and we also know that experience and attention can have an influence on grouping, suggesting that portions of the grouping process are under conscious, cognitive control. The ways in which conscious and unconscious processes work together—and the brain mechanisms that underlie them—are still being debated, but we've come a long way toward understanding them in the last ten years. We've finally gotten to the point where we can pinpoint specific areas of the brain that are involved in particular aspects of music processing. We even think we know which part of the brain causes you to pay attention to things.

How are thoughts formed? Are memories "stored" in a particular part of the brain? Why do songs sometimes get stuck in your head and you can't get them out? Does your brain take some sick pleasure in slowly driving you crazy with inane commercial jingles? I take up these and other ideas in the coming chapters.

3. Behind the Curtain

Music and the Mind Machine

For cognitive scientists, the word *mind* refers to that part of each of us that embodies our thoughts, hopes, desires, memories, beliefs, and experiences. The brain, on the other hand, is an organ of the body, a collection of cells and water, chemicals and blood vessels, that resides in the skull. Activity in the brain gives rise to the contents of the mind. Cognitive scientists sometimes make the analogy that the brain is like a computer's CPU, or hardware, while the mind is like the programs or software running on the CPU. (If only that were literally true and we could just run out to buy a memory upgrade.) Different programs can run on what is essentially the same hardware—different minds can arise from very similar brains.

Western culture has inherited a tradition of dualism from René Descartes, who wrote that the mind and the brain are two entirely separate things. Dualists assert that the mind preexisted, before you were born, and that the brain is not the seat of thought—rather, it is merely an instrument of the mind, helping to implement the mind's will, move muscles, and maintain homeostasis in the body. To most of us, it certainly feels as though our minds are something unique and distinctive, separate from just a bunch of neurochemical processes. We have a feeling of what it is like to be me, what it is like to be me reading a book, and what it is like to think about what it is like to be me. How can *me* be reduced so unceremoniously to axons, dendrites, and ion channels? It feels like we are something more.

But this feeling could be an illusion, just as it certainly feels as though the earth is standing still, not spinning around on its axis at a thousand miles per hour. Most scientists and contemporary philosophers believe that the brain and mind are two parts of the same thing, and some believe that the distinction itself is flawed. The dominant view today is that that the sum total of your thoughts, beliefs, and experiences is represented in patterns of firings—electrochemical activity—in the brain. If the brain ceases to function, the mind is gone, but the brain can still exist, thoughtless, in a jar in someone's laboratory.

Evidence for this comes from neuropsychological findings of regional specificity of function. Sometimes, as a result of stroke (a blockage of blood vessels in the brain that leads to cell death), tumors, head injury, or other trauma, an area of the brain becomes damaged. In many of these cases, damage to a specific brain region leads to a loss of a particular mental or bodily function. When dozens or hundreds of cases show loss of a specific function associated with a particular brain region, we infer that this brain region is somehow involved in, or perhaps responsible for, that function.

More than a century of such neuropsychological investigation has allowed us to make maps of the brain's areas of function, and to localize particular cognitive operations. The prevailing view of the brain is that it is a computational system, and we think of the brain as a type of computer. Networks of interconnected neurons perform computations on information and combine their computations in ways that lead to thoughts, decisions, perceptions, and ultimately consciousness. Different subsystems are responsible for different aspects of cognition. Damage to an area of the brain just above and behind the left ear—Wernicke's area—causes difficulty in understanding spoken language; damage to a region at the very top of the head—the motor cortex—causes difficulty moving your fingers; damage to an area in the center of the brain—the hippocampal complex—can block the ability to form new memories, while leaving old memories intact. Damage to an area just behind your forehead can cause dramatic changes in personality—it can rob aspects of you from you. Such localization of mental function is a strong scientific argument for the involvement of the brain in thought, and the thesis that thoughts come from the brain.

We have known since 1848 (and the medical case of Phineas Gage) that the frontal lobes are intimately related to aspects of self and personality. Yet even one hundred and fifty years later, most of what we can say about personality and neural structures is vague and quite general. We have not located the "patience" region of the brain, nor the "jealousy" or "generous" regions, and it seems unlikely that we ever will. The brain has regional differentiation of structure and function, but complex personality attributes are no doubt distributed widely throughout the brain.

The human brain is divided up into four lobes—the frontal, temporal, parietal, and occipital—plus the cerebellum. We can make some gross generalizations about function, but in fact behavior is complex and not readily reducible to simple mappings. The frontal lobe is associated with planning, and with self-control, and with making sense out of the dense and jumbled signals that our senses receive—the so-called "perceptual organization" that the Gestalt psychologists studied. The temporal lobe is associated with hearing and memory. The parietal lobe is associated with motor movements and spatial skill, and the occipital lobe with vision. The cerebellum is involved in emotions and the planning of movements, and is the evolutionarily oldest part of our brain; even many animals, such as reptiles, that lack the "higher" brain region of the cortex still have a cerebellum. The surgical separation of a portion of the frontal lobe, the prefrontal cortex, from the thalamus is called a lobotomy. So when the Ramones sang "Now I guess I'll have to tell 'em/That I got no cerebellum" in their song "Teenage Lobotomy" (words and music by Douglas Colvin, John Cummings, Thomas Erdely, and Jeffrey Hyman) they were not being anatomically accurate, but for the sake of artistic license, and for creating one of the great rhymes in rock music, it is hard to begrudge them that.

Musical activity involves nearly every region of the brain that we know about, and nearly every neural subsystem. Different aspects of the music are handled by different neural regions—the brain uses functional segregation for music processing, and employs a system of feature detectors whose job it is to analyze specific aspects of the musical signal, such as pitch, tempo, timbre, and so on. Some of the music processing has points in common with the operations required to analyze other sounds; understanding speech, for example, requires that we segment a flurry of sounds into words, sentences, and phrases, and that we be able to understand aspects beyond the words, such as sarcasm (isn't *that* interesting). Several different dimensions of a musical sound need to be analyzed—usually involving several quasi-independent neural processes—and they then need to be brought together to form a coherent representation of what we're listening to.

Listening to music starts with subcortical (below-the-cortex) structures—the cochlear nuclei, the brain stem, the cerebellum—and then moves up to auditory cortices on both sides of the brain. Trying to follow along with music that you know—or at least music in a style you're familiar with, such as baroque or blues—recruits additional regions of the brain, including the hippocampus—our memory center—and subsections of the frontal lobe, particularly a region called inferior frontal cortex, which is in the lowest parts of the frontal lobe, i.e., closer to your chin than to the top of your head. Tapping along with music, either actually or just in your mind, involves the cerebellum's timing circuits. Performing music—regardless of what instrument you play, or whether you sing, or conduct—involves the frontal lobes again for the planning of your behavior, as well as the motor cortex in the parietal lobe just underneath the top of your head, and the sensory cortex, which provides the tactile feedback that you have pressed the right key on your instrument, or moved the baton where you thought you did. Reading music involves the visual cortex, in the back of your head in the occipetal lobe. Listening to or recalling lyrics invokes language centers, including Broca's and Wernicke's area, as well as other language centers in the temporal and frontal lobes.

At a deeper level, the emotions we experience in response to music involve structures deep in the primitive, reptilian regions of the cerebellar vermis, and the amygdala—the heart of emotional processing in the cortex. The idea of regional specificity is evident in this summary but a complementary principle applies as well, that of distribution of function. The brain is a massively parallel device, with operations distributed widely throughout. There is no single language center, nor is there a single music center. Rather, there are regions that peform component operations, and other regions that coordinate the bringing together of this information. Finally, we have discovered only recently that the brain has a capacity for reorganization that vastly exceeds what we thought before. This ability is called neuroplasticity, and in some cases, it suggests that regional specificity may be temporary, as the processing centers for important mental functions actually move to other regions after trauma or brain damage.

It is difficult to appreciate the complexity of the brain because the numbers are so huge they go well beyond our everyday experience (unless you are a cosmologist). The average brain consists of one hundred billion (100,000,000,000) neurons. Suppose each neuron was one dollar, and you stood on a street corner trying to give dollars away to people as they passed by, as fast as you could hand them out—let's say one dollar per second. If you did this twenty-four hours a day, 365 days a year, without stopping, and if you had started on the day that Jesus was born, you would by the present day only have gone through about two thirds of your money. Even if you gave away *hundred*-dollar bills once a second, it would take you thirty-two years to pass them all out. This is a lot of neurons, but the real power and complexity of the brain (and of thought) come through their connections.

Each neuron is connected to other neurons—usually one thousand to ten thousand others. Just four neurons can be connected in sixty-three ways, or not at all, for a total of sixty-four possibilities. As the number of neurons increases, the number of possible connections grows exponentially (the formula for the way that n neurons can be connected to each other is $2^{(n*(n-1)/2)}$):

> For 2 neurons there are 2 possibilities for how they can be connected
> For 3 neurons there are 8 possibilities
> For 4 neurons there are 64 possibilities
> For 5 neurons there are 1,024 possibilities
> For 6 neurons there are 32,768 possibilities

The number of combinations becomes so large that it is unlikely that we will ever understand all the possible connections in the brain, or what they mean. The number of combinations possible—and hence the number of possible different thoughts or brain states each of us can have—exceeds the number of known particles in the entire known universe.

Similarly, you can see how it is that all the songs we have ever heard—and all those that will ever be created—could be made up of just twelve musical notes (ignoring octaves). Each note can go to another note, or to itself, or to a rest, and this yields twelve possibilities. But each of those possibilities yields twelve more. When you factor in rhythm—each note can take on one of many different note lengths—the number of possibilities grows very, very rapidly.

Much of the brain's computational power comes from this enormous possibility for interconnection, and much of it comes from the fact that brains are parallel processing machines, rather than serial processors. A serial processor is like an assembly line, handling each piece of information as it comes down the mental conveyor belt, performing some operation on that piece of information, and then sending it down the line for the next operation. Computers work like this. Ask a computer to download a song from a Web site, tell you the weather in Boise, and save a file you've been working on, and it will do them one at a time; it does things so fast that it can seem as though it is doing them at the same time—in parallel—but it isn't. Brains, on the other hand, can work on many things at once, overlapping and in parallel. Our auditory system processes sound in this way—it doesn't have to wait to find out what the pitch of a sound is to know where it is coming from; the neural circuits devoted to these two operations are trying to come up with answers at the same time. If one neural circuit finishes its work before another, it just sends its information to other connected brain regions and they can begin using it. If late-arriving information that affects an interpretation of what we're hearing comes in from a separate processing circuit, the brain can "change its mind" and update what it thinks is out there. Our brains are updating their opinions all the time—particularly when it comes to perceiving visual and auditory stimuli—hundreds of times per second, and we don't even know it.

Here's an analogy to convey how neurons connect to each other. Imagine that you're sitting home alone one Sunday morning. You don't feel much of one way or another—you're not particularly happy, not particularly sad, neither angry, excited, jealous, nor tense. You feel more or less neutral. You have a bunch of friends, a network of them, and you can call any of them on the phone. Let's say that each of your friends is rather one dimensional and that they can exert a great influence on your mood. You know, for example, that if you telephone your friend Hannah she'll put you in a happy mood. Whenever you talk to Sam it makes you sad, because the two of you had a third friend who died and Sam reminds you of that. Talking to Carla makes you calm and serene, because she has a soothing voice and you're reminded of the times you sat in a beautiful forest clearing with her, soaking up the sun and meditating. Talking to Edward makes you feel energized; talking to Tammy makes you feel tense. You can pick up your telephone and connect to any of these friends and induce a certain emotion.

You might have hundreds or thousands of these one-dimensional friends, each capable of evoking a particular memory, experience, or mood state. These are your connections. Accessing them causes you to change your mood, or state. If you were to talk to Hannah and Sam at the same time, or one right after the other, Hannah would make you feel happy, Sam would make you feel sad, and in the end you'd be back to where you were—neutral. But we can add an additional nuance, which is the weight or force-of-influence of these connections—how close you feel to an individual at a particular point in time. That weight determines the amount of influence the person will have on you. If you feel twice as close to Hannah as you do to Sam, talking to Hannah and Sam for an equal amount of time would still leave you feeling happy, although not as happy as if you had talked to Hannah alone—Sam's sadness brings you down, but only halfway from the happiness you gained from talking to Hannah.

Let's say that all of these people can talk to one another, and in so do-ing, their states can be modified to some extent. Although your friend Hannah is dispositionally cheery, her cheerfulness can be attenuated by a conversation she has with Sad Sam. If you phone Edward the energizer after he's just spoken with Tense Tammy (who has just gotten off the phone with Jealous Justine), Edward may make you feel a new mix of emotions you've never experienced before, a kind of tense jealousy that you have a lot of energy to go out and do something about. And any of these friends might telephone you at any time, evoking these states in you as a complex chain of feelings or experiences that has gone around, each one influencing the other, and you, in turn, will leave your emo-tional mark on them. With thousands of friends interconnected like this, and a bunch of telephones in your living room ringing off the hook all day long, the number of emotional states you might experience would in-deed be quite varied.

It is generally accepted that our thoughts and memories arise from the myriad connections of this sort that our neurons make. Not all neu-rons are equally active at one time, however—this would cause a ca-cophony of images and sensations in our heads (in fact, this is what happens in epilepsy). Certain groups of neurons—we can call them net-works—become active during certain cognitive activities, and they in turn can activate other neurons. When I stub my toe, the sensory recep-tors in my toe send signals up to the sensory cortex in my brain. This sets off a chain of neural activations that causes me to experience pain, withdraw my foot from the object I stubbed it against, and that might cause my mouth to open involuntarily and shout "&%@!"

When I hear a car horn, air molecules impinging on my eardrum cause electrical signals to be sent to my auditory cortex. This causes a cascade of events that recruits a very different group of neurons than toe stubbing. First, neurons in the auditory cortex process the pitch of the sound so that I can distinguish the car horn from something with a different pitch like a truck's air horn, or the air-horn-in-a-can at a football game. A different group of neurons is activated to determine the location from which the sound came. These and other processes invoke a visual orienting response—I turn toward the sound to see what made it, and instantaneously, if necessary, I jump back (the result of activity from the neurons in my motor cortex, orchestrated with neurons in my emotional center, the amygdala, telling me that danger is imminent).

When I hear Rachmaninoff's Piano Concerto no. 3, the hair cells in my cochlea parse the incoming sound into different frequency bands, sending electrical signals to my primary auditory cortex—area A1— telling it what frequencies are present in the signal. Additional regions in the temporal lobe, including the superior temporal sulcus and the superior temporal gyrus on both sides of the brain, help to distinguish the different timbres I'm hearing. If I want to label those timbres, the hippocampus helps to retrieve the memory of similar sounds I've heard before, and then I'll need to access my mental dictionary—which will require using structures found at the junction between the temporal, occipetal, and parietal lobes. So far, these regions are the same ones, although activated in different ways and with different populations of neurons, that I would use to process the car horn. Whole new populations of neurons will become active, however, as I attend to pitch sequences (dorsalateral prefrontal cortex, and Brodmann areas 44 and 47), rhythms (the lateral cerebellum and the cerebellar vermis), and emotion (frontal lobes, cerebellum, the amygdala, and the nucleus accumbens—part of a network of structures involved in feelings of pleasure and reward, whether it is through eating, having sex, or listening to pleasurable music).

To some extent, if the room is vibrating with the deep sounds of the double bass, some of those same neurons that fired when I stubbed my toe may fire now—neurons sensitive to tactile input. If the car horn has a pitch of A440, neurons that are set to fire when that frequency is encountered will most probably fire, and they'll fire again when an A440 occurs in Rachmaninoff. But my inner mental experience is likely to be different because of the different contexts involved and the different neural networks that are recruited in the two cases.

My experience with oboes and violins is different, and the particular way that Rachmaninoff uses them may cause me to have the opposite reaction to his concerto than I have to the car horn; rather than feeling startled, I feel relaxed. The same neurons that fire when I feel calm and safe in my environment may be triggered by the calm parts of the concerto.

Through experience, I've learned to associate car horns with danger, or at least with someone trying to get my attention. How did this happen? Some sounds are intrinsically soothing while others are frightening. Although there is a great deal of interpersonal variation, we are born with a predisposition toward interpreting sounds in particular ways. Abrupt, short, loud sounds tend to be interpreted by many animals as an alert sound; we see this when comparing the alert calls of birds, rodents, and apes. Slow onset, long, and quieter sounds tend to be interpreted as calming, or at least neutral. Think of the sharp sound of a dog's bark, versus the soft purring of a cat who sits peacefully on your lap. Composers know this, of course, and use hundreds of subtle shadings of timbre and note length to convey the many different emotional shadings of human experience.

In the "Surprise Symphony" by Haydn (Symphony no. 94 in G Major, second movement, andante), the composer builds suspense by using soft violins in the main theme. The softness of the sound is soothing, but the shortness of the pizzicato accompaniment sends a gentle, contradictory message of danger, and together they give a soft sense of suspense. The main melodic idea spans barely more than half an octave, a perfect fifth. The melodic contour further suggests complacency—the melody first goes up, then down, then repeats the "up" motif. The parallelism implied by the melody, the up/down/up, gets the listener ready for another "down" part. Continuing with the soft, gentle violin notes, the maestro changes the melody by going up—just a little—but holds the rhythms constant. He rests on the fifth, a relatively stable tone harmonically. Because the fifth is the highest note we've encountered so far, we expect that when the next note comes in, it will be lower—that it will begin the return home toward the root (or tonic), and "close the gap" created by the distance between the tonic and the current note—the fifth. Then, from out of nowhere, Haydn sends us a loud note an octave higher, with the brash horns and timpani carrying the sound. He has violated our expectations for melodic direction, for contour, for timbre, and for loudness all at once. This is the "Surprise" in the "Surprise Symphony."

This Haydn symphony violates our expectations of how the world works. Even someone with no musical knowledge or musical expectations whatsoever finds the symphony surprising because of this timbral effect, switching from the soft purring of the violins to the alert call of horns and drums. For someone with a musical background, the symphony violates expectations that have been formed based on musical convention and style. Where do surprises, expectations, and analyses of this sort occur in the brain? Just how these operations are carried out in neurons is still something of a mystery, but we do have some clues.

Before going any farther, I have to admit a bias in the way I approach the scientific study of minds and brains: I have a definite preference for studying the mind rather than the brain. Part of my preference is personal rather than professional. As a child I wouldn't collect butterflies with the rest of my science class because life—all life—seems sacred to me. And the stark fact about brain research over the course of the last century is that it generally involves poking around in the brains of live animals, often our close genetic cousins, the monkeys and apes, and then killing (they call it "sacrificing") the animal. I worked for one miserable semester in a monkey lab, dissecting the brains of dead monkeys to prepare them for microscopic examination. Every day I had to walk by cages of the ones that were still alive. I had nightmares.

At a different level, I've always been more fascinated by the thoughts themselves, not the neurons that give rise to them. A theory in cognitive science named functionalism—which many prominent researchers subscribe to—asserts that similar minds can arise from quite different brains, that brains are just the collection of wires and processing modules that instantiate thought. Regardless of whether the functionalist doctrine is true, it does suggest that there are limits to how much we can know about thought from just studying brains. A neurosurgeon once told Daniel Dennett (a prominent and persuasive spokesperson for functionalism) that he had operated on hundreds of people and seen hundreds of live, thinking brains, but he had never seen a thought.

When I was trying to decide where to attend graduate school, and who I wanted to have as a mentor, I was infatuated with the work of Professor Michael Posner. He had pioneered a number of ways of looking at thought processes, among them mental chronometry (the idea that much can be learned about the organization of the mind by measuring how long it takes to think certain thoughts), ways to investigate the structure of categories, and the famous Posner Cueing Paradigm, a novel method for studying attention. But rumor had it that Posner was abandoning the mind and had started studying the brain, something I was certain I did not want to do.

Although still an undergraduate (albeit a somewhat older one than usual), I attended the annual meeting of the American Psychological Association, which was held in San Francisco that year, just forty miles up the road from Stanford, where I was finishing up my B.A. I saw Posner's name on the program and attended his talk, which was full of slides containing pictures of people's brains while they were doing one thing or another. After his talk was over he took some questions, then disappeared out a back door. I ran around to the back and saw him way ahead, rushing across the conference center to get to another talk. I ran to catch up to him. I must have been quite a sight to him! I was out of breath from running. Even without the panting, I was nervous meeting one of the great legends of cognitive psychology. I had read his textbook in my first psychology class at MIT (where I began my undergraduate training before transferring to Stanford); my first psychology professor, Susan Carey, spoke of him with what could only be described as reverence in her voice. I can still remember the echoes of her words, reverberating through the lecture hall at MIT: "Michael Posner, one of the smartest and most creative people I've ever met."

I started to sweat, I opened my mouth, and . . . nothing. I started "Mmm . . ." All this time we were walking rapidly side by side—he's a fast walker—and every two or three steps I'd fall behind again. I stammered an introduction and said that I had applied to the University of Oregon to work with him. I'd never stuttered before, but I had never been this nervous before. "P-p-p-professor P-p-posner, I hear that you've shifted your research focus entirely to the b-b-brain—is that true? Because I really want to study cognitive psychology with you," I finally told him.

"Well, I am a little interested in the brain these days," he said. "But I see cognitive neuroscience as a way to provide constraints for our theories in cognitive psychology. It helps us to distinguish whether a model has a plausible basis in the underlying anatomy."

Many people enter neuroscience from a background in biology or chemistry and their principal focus is on the mechanisms by which cells communicate with each other. To the cognitive neuroscientist, understanding the anatomy or physiology of the brain may be a challenging intellectual exercise (the brain scientists' equivalent of a really complicated crossword puzzle), but it is not the ultimate goal of the work. Our goal is to understand thought processes, memories, emotions, and experiences, and the brain just happens to be the box that all this happens in. To return to the telephone analogy and conversations you might have with different friends who influence your emotions: If I want to predict how you're going to feel tomorrow, it will be of only limited value for me to map the layout of the telephone lines connecting all the different people you know. More important is to understand their individual proclivities: Who is likely to call you tomorrow and what are they likely to say? How are they apt to make you feel? Of course, to entirely ignore the connectivity question would be a mistake too. If a line is broken, or if there is no evidence of a connection between person A and person B, or if person C can never call you directly but can only influence you through person A who can call you directly—all this information provides important constraints to a prediction.

This perspective influences the way I study the cognitive neuroscience of music. I am not interested in going on a fishing expedition to try every possible musical stimulus and find out where it occurs in the brain; Posner and I have talked many times about the current mad rush to map the brain as just so much atheoretical cartography. The point for me isn't to develop a map of the brain, but to understand how it works, how the different regions coordinate their activity together, how the simple firing of neurons and shuttling around of neurotransmitters leads to thoughts, laughter, feelings of profound joy and sadness, and how all these, in turn, can lead us to create lasting, meaningful works of art. These are the functions of the mind, and knowing where they occur doesn't interest me unless the where can tell us something about how and why. An assumption of cognitive neuroscience is that it can.

My perspective is that, of the infinite number of experiments that are possible to do, the ones worth doing are those that can lead us to a better understanding of how and why. A good experiment is theoretically motivated, and makes clear predictions as to which one of two or more competing hypotheses will be supported. An experiment that is likely to provide support for both sides of a contentious issue is not one worth doing; science can only move forward by the elimination of false or untenable hypotheses.

Another quality of a good experiment is that it is generalizable to other conditions—to people not studied, to types of music not studied, and to a variety of situations. A great deal of behavioral research is conducted on only a small number of people ("subjects" in the experiment), and with very artificial stimuli. In my laboratory we use both musicians and nonmusicians whenever possible, in order to learn about the broadest cross section of people. And we almost always use real-world music, actual recordings of real musicians playing real songs, so that we can better understand the brain's responses to the kind of music that most people listen to, rather than the kind of music that is found only in the neuroscientific laboratory. So far this approach has panned out. It is more difficult to provide rigorous experimental controls with this approach, but it is not impossible; it takes a bit more planning and careful preparation, but in the long run, the results are worth it. In using this naturalistic approach, I can state with reasonable scientific certainty that we're studying the brain doing what it normally does, rather than what it does when assaulted by rhythms without any pitch, or melodies without any rhythms. In an attempt to separate music into its components, we run the risk—if the experiments are not done properly—of creating sound sequences that are very unmusical.

When I say that I am less interested in the brain than in the mind, this does not mean that I have no interest in the brain. I believe that we all have brains, and I believe brains are important! But I also believe similar thoughts can arise from different brain architectures. By analogy, I can watch the same television program on an RCA, a Zenith, a Mitsubishi, even on my computer screen with the right hardware and software. The architectures of all these are sufficiently distinct from one another that the patent office—an organization charged with the responsibility of deciding when something is sufficiently different from something else that it constitutes an invention—has issued different patents to these various companies, establishing that the underlying architectures are significantly different. My dog Shadow has a very different brain organization, anatomy, and neurochemistry from mine. When he is hungry or hurts his paw, it is unlikely that the pattern of nerve firings in his brain bears much resemblance to the pattern of firings in my brain when I'm hungry or stub my toe. But I do believe that he is experiencing substantially similar mind states.

Some common illusions and misconceptions need to be set aside. Many people, even trained scientists in other disciplines, have the strong intuition that inside the brain there is a strictly isomorphic representation of the world around us. (*Isomorphic* comes from the Greek word *iso*, meaning "same," and *morphus*, meaning "form.") The Gestalt psychologists, who were right about a great many things, were among the first to articulate this idea. If you look at a square, they argued, a square-shaped pattern of neurons is activated in your brain. Many of us have the intuition that if we're looking at a tree, the image of the tree is somewhere represented in the brain as a tree, and that perhaps seeing the tree activates a set of neurons in the shape of a tree, with roots at one end and leaves at the other. When we listen to or imagine a favorite song, it feels like the song is playing in our head, over a set of neural loudspeakers.

Daniel Dennett and V. S. Ramachandran have eloquently argued that there is a problem with this intuition. If a mental picture of something (either as we see it right now or imagine it in memory) is itself a picture, there has to be some part of our mind/brain that is seeing that picture. Dennett talks about the intuition that visual scenes are presented on some sort of a screen or theater in our minds. For this to be true, there would have to be someone in the audience of that theater watching the screen, and holding a mental image inside his head. And who would that be? What would that mental image look like? This quickly leads to an infinite regress. The same argument applies to auditory events. No one argues that it doesn't feel like we have an audio system in our minds. Because we can manipulate mental images—we can zoom in on them, rotate them, in the case of music we can speed up or slow down the song in our heads—we're compelled to think there is a home theater in the mind. But logically this cannot be true because of the infinite regress problem.

We are also under the illusion that we simply open our eyes and—we see. A bird chirps outside the window and we instantly hear. Sensory perception creates mental images in our minds—representations of the world outside our heads—so quickly and seamlessly that it seems there is nothing to it. This is an illusion. Our perceptions are the end product of a long chain of neural events that give us the illusion of an instantaneous image. There are many domains in which our strongest intuitions mislead us. The flat earth is one example. The intuition that our senses give us an undistorted view of the world is another.

It has been known at least since the time of Aristotle that our senses can distort the way we perceive the world. My teacher Roger Shepard, a perception psychologist at Stanford University, used to say that when functioning properly, our perceptual system is supposed to distort the world we see and hear. We interact with the world around us through our senses. As John Locke noted, everything we know about the world is through what we see, hear, smell, touch, or taste. We naturally assume that the world is just as we perceive it to be. But experiments have forced us to confront the reality that this is not the case. Visual illusions are perhaps the most compelling proof of sensory distortion. Many of us have seen these sorts of illusions as children, such as when two lines of the same length appear to be different lengths (the Ponzo illusion).

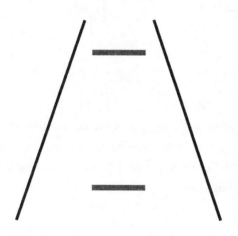

Roger Shepard drew an illusion he calls "Turning the Tables" that is related to the Ponzo. It's hard to believe, but these tabletops are identical in size and shape (you can check by cutting out a piece of paper or cellophane the exact shape of one and then placing it over the other). This illusion exploits a principle of our visual system's depth perception mechanisms. Even knowing that it is an illusion does not allow us to turn off the mechanism. No matter how many times we view this figure, it continues to surprise us because our brains are actually giving us misinformation about the objects.

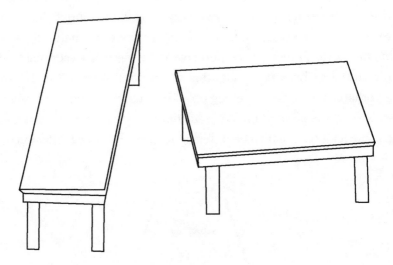

In the Kaniza illusion there appears to be a white triangle lying on top of a black-outlined one. But if you look closely, you'll see that there are no triangles in the figure. Our perceptual system completes or "fills in" information that isn't there.

Why does it do this? Our best guess is that it was evolutionarily adaptive to do so. Much of what we see and hear contains missing information. Our hunter-gatherer ancestors might have seen a tiger partially hidden by trees, or heard a lion's roar partly obscured by the sound of leaves rustling much closer to us. Sounds and sights often come to us as partial information that has been obscured by other things in the environment. A perceptual system that can restore missing information would help us make quick decisions in threatening situations. Better to run now than sit and try to figure out if those two separate, broken pieces of sound were part of a single lion roar.

The auditory system has its own version of perceptual completion. The cognitive psychologist Richard Warren demonstrated this particularly well. He recorded a sentence, "The bill was passed by both houses of the legislature," and cut out a piece of the sentence from the recording tape. He replaced the missing piece with a burst of white noise (static) of the same duration. Nearly everyone who heard the altered recording could report that they heard both a sentence and static. But a large proportion of people couldn't tell where the static was! The auditory system had filled in the missing speech information, so that the sentence seemed to be uninterrupted. Most people reported that there was static and that it existed apart from the spoken sentence. The static and the sentence formed separate perceptual streams due to differences in timbre that caused them to group separately; Bregman calls this streaming by timbre. Clearly this is a sensory distortion; our perceptual system is telling us something about the world that isn't true. But just as clearly, this has an evolutionary/adaptive value if it can help us make sense of the world during a life-or-death situation.

According to the great perception psychologists Hermann von Helmholtz, Richard Gregory, Irvin Rock, and Roger Shepard, perception is a process of inference, and involves an analysis of probabilities. The brain's task is to determine what the most likely arrangement of objects in the physical world is, given the particular pattern of information that reaches the sensory receptors—the retina for vision, the eardrum for hearing. Most of the time the information we receive at our sensory receptors is incomplete or ambiguous. Voices are mixed in with other voices, the sounds of machines, wind, footsteps. Wherever you are right now—whether you're in an airplane, a coffee shop, a library, at home, in a park, or anywhere else—stop and listen to the sounds around you. Unless you're in a sensory isolation tank, you can probably identify at least a half-dozen different sounds. Your brain's ability to make these identifications is nothing short of remarkable when you consider what it starts out with—that is, what the sensory receptors pass up to it. Grouping principles—by timbre, spatial location, loudness, and so on—help to segregate them, but there is still a lot we don't know about this process; no one has yet designed a computer that can perform this task of sound source separation.

The eardrum is simply a membrane that is stretched across tissue and bone. It is the gateway to hearing. Virtually all of your impressions of the auditory world come from the way in which it wiggles back and forth in response to air molecules hitting it. (To a degree, the pinnae—the fleshy parts of your ear—are also involved in auditory perception, as are the bones in your skull, but for the most part, the eardrum is the primary source of what we know about what is out there in the auditory world.) Let's consider a typical auditory scene, a person sitting in her living room reading a book. In this environment, let's suppose that there are six sources of sound that she can readily identify: the whooshing noise of the central heating (the fan or blower that moves air through the ductwork), the hum of a refrigerator in the kitchen, traffic outside on the street (which itself could be several or dozens of distinct sounds comprising different engines, brakes squeaking, horns, etc.), leaves rustling in the wind outside, a cat purring on the chair next to her, and a recording of Debussy preludes. Each of these can be considered an auditory object or a sound source, and we are able to identify them because each has its own distinctive sound.

Sound is transmitted through the air by molecules vibrating at certain frequencies. These molecules bombard the eardrum, causing it to wiggle in and out depending on how hard they hit it (related to the volume or amplitude of the sound) and on how fast they're vibrating (related to what we call pitch). But there is nothing in the molecules that tells the eardrum where they came from, or which ones are associated with which object. The molecules that were set in motion by the cat purring don't carry an identifying tag that says cat, and they may arrive on the eardrum at the same time and in the same region of the eardrum as the sounds from the refrigerator, the heater, Debussy, and everything else.

Imagine that you stretch a pillowcase tightly across the opening of a bucket, and different people throw Ping-Pong balls at it from different distances. Each person can throw as many Ping-Pong balls as he likes, and as often as he likes. Your job is to figure out, just by looking at how the pillowcase moves up and down, how many people there are, who they are, and whether they are walking toward you, away from you, or are standing still. This is analogous to what the auditory system has to contend with in making identifications of auditory objects in the world, using only the movement of the eardrum as a guide. How does the brain figure out, from this disorganized mixture of molecules beating against a membrane, what is out there in the world? In particular, how does it do this with music?

It does this through a process of feature extraction, followed by another process of feature integration. The brain extracts basic, low-level features from the music, using specialized neural networks that decompose the signal into information about pitch, timbre, spatial location, loudness, reverberant environment, tone durations, and the onset times for different notes (and for different components of tones). These operations are carried out in parallel by neural circuits that compute these values and that can operate somewhat independently of one another— that is, the pitch circuit doesn't need to wait for the duration circuit to be done in order to perform its calculations. This sort of processing— where only the information contained in the stimulus is considered by the neural circuits—is called bottom-up processing. In the world and in the brain, these attributes of the music are separable. We can change one without changing the other, just as we can change shape in visual objects without changing their color.

Low-level, bottom-up processing of basic elements occurs in the peripheral and phylogenetically older parts of our brains; the term *low-level* refers to the perception of elemental or building-block attributes of a sensory stimulus. High-level processing occurs in more sophisticated parts of our brains that take neural projections from the sensory receptors and from a number of low-level processing units; this refers to the combining of low-level elements into an integrated representation. High-level processing is where it all comes together, where our minds come to an understanding of form and content. Low-level processing in your brain sees blobs of ink on this page, and perhaps even allows you to put those blobs together and recognize a basic form in your visual vocabulary, such as the letter *A*. But it is high-level processing that puts together three letters to let you read the word ART and to generate a mental image of what the word means.

At the same time as feature extraction is taking place in the cochlea, auditory cortex, brain stem, and cerebellum, the higher-level centers of our brain are receiving a constant flow of information about what has been extracted so far; this information is continually updated, and typically rewrites the older information. As our centers for higher thought—mostly in the frontal cortex—receive these updates, they are working hard to predict what will come next in the music, based on several factors:

~ what has already come before in the piece of music we're hearing;

~ what we remember will come next if the music is familiar;

~ what we expect will come next if the genre or style is familiar, based on previous exposure to this style of music;

~ any additional information we've been given, such as a summary of the music that we've read, a sudden movement by a performer, or a nudge by the person sitting next to us.

These frontal-lobe calculations are called top-down processing and they can exert influence on the lower-level modules while they are performing their bottom-up computations. The top-down expectations can cause us to misperceive things by resetting some of the circuitry in the bottom-up processors. This is partly the neural basis for perceptual completion and other illusions.

The top-down and bottom-up processes inform each other in an ongoing fashion. At the same time as features are being analyzed individually, parts of the brain that are higher up—that is, that are more phylogenetically advanced, and that receive connections from lower brain regions—are working to integrate these features into a perceptual whole. The brain constructs a representation of reality, based on these component features, much as a child constructs a fort out of Lego blocks. In the process, the brain makes a number of inferences, due to incomplete or ambiguous information; sometimes these inferences turn out to be wrong, and that is what visual and auditory illusions are: demonstrations that our perceptual system has guessed incorrectly about what is out-there-in-the-world.

The brain faces three difficulties in trying to identify the auditory objects we hear. First, the information arriving at the sensory receptors is undifferentiated. Second, the information is ambiguous—different objects can give rise to similar or identical patterns of activation on the eardrum. Third, the information is seldom complete. Parts of the sound may be covered up by other sounds, or lost. The brain has to make a calculated guess about what is really out there. It does so very quickly and generally subconsciously. The illusions we saw previously, along with these perceptual operations, are not subject to our awareness. I can tell you, for example, that the reason you see triangles where there are none in the Kaniza figure is due to perceptual completion. But even after you know the principles that are involved, it is impossible to turn them off. Your brain keeps on processing the information in the same way, and you continue to be surprised by the outcome.

Helmholtz called this process "unconscious inference." Rock called it "the logic of perception." George Miller, Ulrich Neisser, Herbert Simon, and Roger Shepard have described perception as a "constructive process." These are all ways of saying that what we see and hear is the end of a long chain of mental events that give rise to an impression, a mental image, of the physical world. Many of the ways in which our brains function—including our senses of color, taste, smell, and hearing—arose due to evolutionary pressures, some of which no longer exist. The cognitive psychologist Steven Pinker and others have suggested that our music-perception system was essentially an evolutionary accident, and that survival and sexual-selection pressures created a language and communication system that we learned to exploit for musical purposes. This is a contentious point in the cognitive-psychology community. The archaeological record has left us some clues, but it rarely leaves us a "smoking gun" that can settle such issues definitively. The filling-in phenomenon I've described is not just a laboratory curiosity; composers exploit this principle as well, knowing that our perception of a melodic line will continue, even if part of it is obscured by other instruments. Whenever we hear the lowest notes on the piano or double bass, we are not actually hearing 27.5 or 35 Hz, because those instruments are typically incapable of producing much energy at these ultralow frequencies: Our ears are filling in the information and giving us the illusion that the tone is that low.

We experience illusions in other ways in music. In piano works such as Sindig's "The Rustle of Spring" or Chopin's Fantasy-Impromptu in C-sharp Minor, op. 66, the notes go by so quickly that an illusory melody emerges. Play the tune slowly and it disappears. Due to stream segregation, the melody "pops out" when the notes are close enough together in time—the perceptual system holds the notes together—but the melody is lost when its notes are too far apart in time. As studied by Bernard Lortat-Jacob at the Musée de l'Homme in Paris, the Quintina (literally "fifth one") in Sardinian a capella vocal music also conveys an illusion: A fifth female voice emerges from the four male voices when the harmony and timbres are performed just right. (They believe the voice is that of the Virgin Mary coming to reward them if they are pious enough to sing it right.)

In the Eagles' "One of These Nights" (the title song from the album of the same name) the song opens with a pattern played by bass and guitar that sounds like one instrument—the bass plays a single note, and the guitar adds a glissando, but the perceptual effect is of the bass sliding, due to the Gestalt principle of good continuation. George Shearing created a new timbral effect by having guitar (or in some cases, vibrophone) double what he was playing on the piano so precisely that listeners come away wondering, "What is that new instrument?" when in reality it is two separate instruments whose sounds have perceptually fused. In "Lady Madonna," the four Beatles sing into their cupped hands during an instrumental break and we swear that there are saxophones playing, based on the unusual timbre they achieve coupled with our (top-down) expectation that saxophones should be playing in a song of this genre.

Most contemporary recordings are filled with another type of auditory illusion. Artificial reverberation makes vocalists and lead guitars sound like they're coming from the back of a concert hall, even when we're listening in headphones and the sound is coming from an inch away from our ears. Microphone techniques can make a guitar sound like it is ten feet wide and your ears are right where the soundhole is— an impossibility in the real world (because the strings have to go across the soundhole—and if your ears were really there, the guitarist would be strumming your nose). Our brains use cues about the spectrum of the sound and the type of echoes to tell us about the auditory world around us, much as a mouse uses his whiskers to know about the physical world around him. Recording engineers have learned to mimic those cues to imbue recordings with a real-world, lifelike quality even when they're made in sterile recording studios.

There is a related reason why so many of us are attracted to recorded music these days—and especially now that personal music players are common and people are listening in headphones a lot. Recording engineers and musicians have learned to create special effects that tickle our brains by exploiting neural circuits that evolved to discern important features of our auditory environment. These special effects are similar in principle to 3-D art, motion pictures, or visual illusions, none of which have been around long enough for our brains to have evolved special mechanisms to perceive them; rather, they leverage perceptual systems that are in place to accomplish other things. Because they use these neural circuits in novel ways, we find them especially interesting. The same is true of the way that modern recordings are made.

Our brains can estimate the size of an enclosed space on the basis of the reverberation and echo present in the signal that hits our ears. Even though few of us understand the equations necessary to describe how one room differs from another, all of us can tell whether we're standing in a small, tiled bathroom, a medium-sized concert hall, or a large church with high ceilings. And we can tell when we hear recordings of voices what size room the singer or speaker is in. Recording engineers create what I call "hyperrealities," the recorded equivalent of the cinematographer's trick of mounting a camera on the bumper of a speeding car. We experience sensory impressions that we never actually have in the real world.

Our brains are exquisitely sensitive to timing information. We are able to localize objects in the world based on differences of only a few milliseconds between the time of arrival of a sound at one of our ears versus the other. Many of the special effects we love to hear in recorded music are based on this sensitivity. The guitar sound of Pat Metheny or David Gilmour of Pink Floyd use multiple delays of the signal to give an otherwordly, haunting effect that triggers parts of our brains in ways that humans had never experienced before, by simulating the sound of an enclosed cave with multiple echoes such as would never actually occur in the real world—an auditory equivalent of the barbershop mirrors that repeated infinitely.

Perhaps the ultimate illusion in music is the illusion of structure and form. There is nothing in a sequence of notes themselves that creates the rich emotional associations we have with music, nothing about a scale, a chord, or a chord sequence that intrinsically causes us to expect a resolution. Our ability to make sense of music depends on experience, and on neural structures that can learn and modify themselves with each new song we hear, and with each new listening to an old song. Our brains learn a kind of musical grammar that is specific to the music of our culture, just as we learn to speak the language of our culture.

Noam Chomsky's contribution to modern linguistics and psychology was proposing that we are all born with an innate capacity to understand any of the world's languages, and that experience with a particular language shapes, builds, and then ultimately prunes a complicated and interconnected network of neural circuits. Our brain doesn't know before we're born which language we'll be exposed to, but our brains and natural languages coevolved so that all of the world's languages share certain fundamental principles, and our brains have the capacity to incorporate any of them, almost effortlessly, through mere exposure during a critical stage of neural development.

Similarly, it seems that we all have an innate capacity to learn any of the world's musics, although they, too, differ in substantive ways from one another. The brain undergoes a period of rapid neural development after birth, continuing for the first years of life. During this time, new neural connections are forming more rapidly than at any other time in our lives, and during our midchildhood years, the brain starts to prune these connections, retaining only the most important and most often used ones. This becomes the basis for our understanding of music, and ultimately the basis for what we like in music, what music moves us, and how it moves us. This is not to say that we can't learn to appreciate new music as adults, but basic structural elements are incorporated into the very wiring of our brains when we listen to music early in our lives.

Music, then, can be thought of as a type of perceptual illusion in which our brain imposes structure and order on a sequence of sounds. Just how this structure leads us to experience emotional reactions is part of the mystery of music. After all, we don't get all weepy eyed when we experience other kinds of structure in our lives, such as a balanced checkbook or the orderly arrangement of first-aid products in a drugstore (well, at least most of us don't). What is it about the particular kind of order we find in music that moves us so? The structure of scales and chords has something to do with it, as does the structure of our brains. Feature detectors in our brains work to extract information from the stream of sounds that hits our ears. The brain's computational system combines these into a coherent whole, based in part on what it thinks it ought to be hearing, and in part based on expectations. Just where those expectations come from is one of the keys to understanding how music moves, when it moves us, and why some music only makes us want to reach for the off button on our radios or CD players. The topic of musical expectations is perhaps the area in the cognitive neuroscience of music that most harmoniously unites music theory and neural theory, musicians and scientists, and to understand it completely, we have to study how particular patterns of music give rise to particular patterns of neural activations in the brain.

4. Anticipation

What We Expect from Liszt (and Ludacris)

When I'm at a wedding, it is not the sight of the hope and love of the bride and groom standing in front of their friends and family, their whole life before them, that makes my eyes tear up. It is when the music begins that I start to cry. In a movie, when two people are at long last reunited after some great ordeal, the music again pushes me and my emotions over the sentimental edge.

I said earlier that music is organized sound, but the organization has to involve some element of the unexpected or it is emotionally flat and robotic. The appreciation we have for music is intimately related to our ability to learn the underlying structure of the music we like—the equivalent to grammar in spoken or signed languages—and to be able to make predictions about what will come next. Composers imbue music with emotion by knowing what our expectations are and then very deliberately controlling when those expectations will be met, and when they won't. The thrills, chills, and tears we experience from music are the result of having our expectations artfully manipulated by a skilled composer and the musicians who interpret that music.

Perhaps the most documented illusion—or parlor trick—in Western classical music is the deceptive cadence. A cadence is a chord sequence that sets up a clear expectation and then closes, typically with a satisfying resolution. In the deceptive cadence, the composer repeats the chord sequence again and again until he has finally convinced the listeners that we're going to get what we expect, but then at the last minute, he gives us an unexpected chord—not outside the key, but a chord that tells us that it's not over, a chord that doesn't completely resolve. Haydn's use of the deceptive cadence is so frequent, it borders on an obsession. Perry Cook has likened this to a magic trick: Magicians set up expectations and then defy them, all without you knowing exactly how or when they're going to do it. Composers do the same thing. The Beatles' "For No One" ends on the V chord (the fifth degree of the scale we're in) and we wait for a resolution that never comes—at least not in that song. But the very next song on the album *Revolver* starts with the very chord we were waiting to hear.

The setting up and then manipulating of expectations is the heart of music, and it is accomplished in countless ways. Steely Dan do it by playing songs that are essentially the blues (with blues structure and chord progressions) but by adding unusual harmonies to the chords that make them sound very unblues—for example on their song "Chain Lightning." Miles Davis and John Coltrane made careers out of reharmonizing blues progressions to give them new sounds that were anchored partly in the familiar and partly in the exotic. On his solo album *Kamakiriad*, Donald Fagen (of Steely Dan) has songs with blues/funk rhythms that lead us to expect the standard blues chord progression, but the entire song is played on only one chord, never moving from that harmonic position.

In "Yesterday," the main melodic phrase is seven measures long; the Beatles surprise us by violating one of the most basic assumptions of popular music, the four- or eight-measure phrase unit (nearly all rock/pop songs have musical ideas that are organized into phrases of those lengths). In "I Want You (She's So Heavy)," the Beatles violate expectations by first setting up a hypnotic, repetitive ending that sounds like it will go on forever; based on our experience with rock music and rock music endings, we expect that the song will slowly die down in volume, the classic fade-out. Instead, they end the song abruptly, and not even at the end of a phrase—they end right in the middle of a note!

The Carpenters use timbre to violate genre expectations; they were probably the last group people expected to use a distorted electric guitar, but they did on "Please Mr. Postman" and some other songs. The Rolling Stones—one of the hardest rock bands in the world at the time—had done the opposite of this just a few years before by using violins (as for example, on "Lady Jane"). When Van Halen were the newest, hippest group around they surprised fans by launching into a heavy metal version of an old not-quite-hip song by the Kinks, "You Really Got Me."

Rhythm expectations are violated often as well. A standard trick in electric blues is for the band to build up momentum and then stop playing altogether while the singer or lead guitarist continues on, as in Stevie Ray Vaughan's "Pride and Joy," Elvis Presley's "Hound Dog," or the Allman Brothers' "One Way Out." The classic ending to an electric blues song is another example. The song charges along with a steady beat for two or three minutes and—wham! Just as the chords suggest an ending is imminent, rather than charging through at full speed, the band suddenly starts playing at half the tempo they were before.

In a double whammy, Creedence Clearwater Revival pulls out this slowed-down ending in "Lookin' Out My Back Door"—by then such an ending was already a well-known cliché—and they violate the expectations of *that* by coming in again for the real ending of the song at full tempo.

The Police made a career out of violating rhythmic expectations. The standard rhythmic convention in rock is to have a strong backbeat on beats two and four. Reggae music turns this around putting the snare drum on beats one and two, and (typically) a guitar on two and four. The Police combined reggae with rock to create a new sound that fulfilled some and violated other rhythmic expectations simultaneously. Sting often played bass guitar parts that were entirely novel, avoiding the rock clichés of playing on the downbeat or of playing synchronously with the bass drum. As Randy Jackson of *American Idol* fame, and one of the top session bass players, told me (back when we shared an office in a recording studio in the 1980s), Sting's basslines are unlike anyone else's, and they wouldn't even fit in anyone else's songs. "Spirits in the Material World" from their album *Ghost in the Machine* takes this rhythmic play to such an extreme it can be hard to tell where the downbeat even is.

Modern composers such as Schönberg threw out the whole idea of expectation. The scales they used deprive us of the notion of a resolution, a root to the scale, or a musical "home," thus creating the illusion of no home, a music adrift, perhaps as a metaphor for a twentieth-century existentialist existence (or just because they were trying to be contrary). We still hear these scales used in movies to accompany dream sequences to convey a lack of grounding, or in underwater or outer space scenes to convey weightlessness.

These aspects of music are not represented directly in the brain, at least not during initial stages of processing. The brain constructs its own version of reality, based only in part on what is there, and in part on how it interprets the tones we hear as a function of the role they play in a learned musical system. We interpret spoken language analogously. There is nothing intrinsically catlike about the word *cat* or even any of its syllables. We have learned that this collection of sounds represents the feline house pet. Similarly, we have learned that certain sequences of tones go together, and we expect them to continue to do so. We expect certain pitches, rhythms, timbres, and so on to co-occur based on a statistical analysis our brain has performed of how often they have gone together in the past. We have to reject the intuitively appealing idea that the brain is storing an accurate and strictly isomorphic representation of the world. To some degree, it is storing perceptual distortions, illusions, and extracting relationships among elements. It is computing a reality for us, one that is rich in complexity and beauty. A basic piece of evidence for such a view is the simple fact that light waves in the world vary along one dimension—wavelength—and yet our perceptual system treats color as two dimensional (the color circle described on page 29). Similarly with pitch: From a one-dimensional continuum of molecules vibrating at different speeds, our brains construct a rich, multidimensional pitch space with three, four, or even five dimensions (according to some models). If our brain is adding this many dimensions to what is out there in the world, this can help explain the deep reactions we have to sounds that are properly constructed and skillfully combined.

When cognitive scientists talk about expectations and violating them, we mean an event whose occurrence is at odds with what might have been reasonably predicted. It is clear that we know a great deal about a number of different standard situations. Life presents us with similar situations that differ only in details, and often those details are insignificant. Learning to read is an example. The feature extractors in our brain have learned to detect the essential and unvarying aspect of letters of the alphabet, and unless we explicitly pay attention, we don't notice details such as the font that a word is typed in. Even though *surface* details are **different**, *all* these words **are** equally ***recognizable***, as **are** their *individual* ***letters***. (It may be jarring to read sentences in which every word is in a different font, and of course such rapid shifting causes us to notice, but the point remains that our feature detectors are busy extracting things like "the letter *a*" rather than processing the font it is typed in.)

An important way that our brain deals with standard situations is that it extracts those elements that are common to multiple situations and creates a framework within which to place them; this framework is called a schema. The schema for the letter *a* would be a description of its shape, and perhaps a set of memory traces that includes all the *a*'s we've ever seen, showing the variability that accompanies the schema. Schemas inform a host of day-to-day interactions we have with the world. For example, we've been to birthday parties and we have a general notion—a schema—of what is common to birthday parties. The birthday party schema will be different for different cultures (as is music), and for people of different ages. The schema leads to clear expectations, as well as a sense of which of those expectations are flexible and which are not. We can make a list of things we would expect to find at a typical birthday party. We wouldn't be surprised if these weren't all present, but the more of them that are absent, the less typical the party would be:

~ A person who is celebrating the anniversary of their birth

~ Other people helping that person to celebrate

~ A cake with candles

~ Presents

~ Festive food

~ Party hats, noisemakers, and other decorations

If the party was for an eight-year-old we might have the additional expectation that there would be a rousing game of pin-the-tail-on-the-donkey, but not single-malt scotch. This more or less constitutes our birthday party schema.

We have musical schemas, too, and these begin forming in the womb and are elaborated, amended, and otherwise informed every time we listen to music. Our musical schema for Western music includes implicit knowledge of the scales that are normally used. This is why Indian or Pakistani music, for example, sounds "strange" to us the first time we hear it. It doesn't sound strange to Indians and Pakistanis, and it doesn't sound strange to infants (or at least not any stranger than any other music). This may be an obvious point, but it sounds strange by virtue of its being inconsistent with what we have learned to call music. By the age of five, infants have learned to recognize chord progressions in the music of their culture—they are forming schemas.

We develop schemas for particular musical genres and styles; *style* is just another word for "repetition." Our schema for a Lawrence Welk concert includes accordions, but not distorted electric guitars, and our schema for a Metallica concert is the opposite. A schema for Dixieland includes foot-tapping, up-tempo music, and unless the band was trying to be ironic, we would not expect there to be overlap between their repertoire and that of a funeral procession. Schemas are an extension of memory. As listeners, we recognize when we are hearing something we've heard before, and we can distinguish whether we heard it earlier in the same piece, or in a different piece. Music listening requires, according to the theorist Eugene Narmour, that we be able to hold in memory a knowledge of those notes that have just gone by, alongside a knowledge of all other musics we are familiar with that approximate the style of what we're listening to now. This latter memory may not have the same level of resolution or the same amount of vividness as notes we've just heard, but it is necessary in order to establish a context for the notes we're hearing.

The principal schemas we develop include a vocabulary of genres and styles, as well as of eras (1970s music sounds different from 1930s music), rhythms, chord progressions, phrase structure (how many measures to a phrase), how long a song is, and what notes typically follow what. When I said earlier that the standard popular song has phrases that are four or eight measures long, this is a part of the schema we've developed for late twentieth-century popular songs. We've heard thousands of songs thousands of times and even without being able to explicitly describe it, we have incorporated this phrase tendency as a "rule" about music we know. When "Yesterday" plays with its seven-measure phrase, it is a surprise. Even though we've heard "Yesterday" a thousand or even ten thousand times, it still interests us because it violates schematic expectations that are even more firmly entrenched than our memory for this particular song. Songs that we keep coming back to for years play around with expectations just enough that they are always at least a little bit surprising. Steely Dan, the Beatles, Rachmaninoff, and Miles Davis are just a few of the artists that some people say they never tire of, and this is a big part of the reason.

Melody is one of the primary ways that our expectations are controlled by composers. Music theorists have identified a principle called gap fill; in a sequence of tones, if a melody makes a large leap, either up or down, the next note should change direction. A typical melody includes a lot of stepwise motion, that is, adjacent tones in the scale. If the melody makes a big leap, theorists describe a tendency for the melody to "want" to return to the jumping-off point; this is another way to say that our brains expect that the leap was only temporary, and tones that follow need to bring us closer and closer to our starting point, or harmonic "home."

In "Somewhere Over the Rainbow," the melody begins with one of the largest leaps we've ever experienced in a lifetime of music listening: an octave. This is a strong schematic violation, and so the composer rewards and soothes us by bringing the melody back toward home again, but not by too much—he does come down, but only by one scale degree—because he wants to continue to build tension. The third note of this melody fills the gap. Sting does the same thing in "Roxanne": He leaps up an interval of roughly a half octave (a perfect fourth) to hit the first syllable of the word *Roxanne*, and then comes down again to fill the gap.

We also hear gap fill in the *andante cantabile* from Beethoven's "Pathétique" Sonata. As the main theme climbs upward, it moves from a C (in the key of A-flat, this is the third degree of the scale) to the A-flat that is an octave above what we consider the "home" note, and it keeps on climbing to a B-flat. Now that we're an octave and a whole step higher than home, there is only one way to go, back toward home. Beethoven actually jumps toward home, down an interval of a fifth, landing on the note (E-flat) that is a fifth above the tonic. To delay the resolution—Beethoven was a master of suspense—instead of continuing the descent down to the tonic, Beethoven moves away from it. In writing the jump down from the high B-flat to the E-flat, Beethoven was pitting two schemas against each other: the schema for resolving to the tonic, and the schema for gap fill. By moving away from the tonic at this point, he is also filling the gap he made by jumping so far down to get to this midpoint. When Beethoven finally brings us home two measures later, it is as sweet a resolution as we've ever heard.

Consider now what Beethoven does to expectations with the melody to the main theme from the last movement of his Ninth Symphony ("Ode to Joy"). These are the notes of the melody, as *solfège*, the do-re-mi system:

mi - mi - fa - sol - sol - fa - me - re - do - do -re - mi - mi - re - re

(If you're having trouble following along, it will help if you sing in your mind the English words to this part of the song: "Come and sing a song of joy for peace a glory gloria . . .")

The main melodic theme is simply the notes of the scale! The best-known, overheard, and overused sequence of notes we have in Western music. But Beethoven makes it interesting by violating our expectations. He starts on a strange note and ends on a strange note. He starts on the third degree of the scale (as he did on the "Pathétique" Sonata), rather than the root, and then goes up in stepwise fashion, then turns around and comes down again. When he gets to the root—the most stable tone—rather than staying there he comes up again, up to the note we started on, then back down so that we think and we expect he will hit the root again, but he doesn't; he stays right there on *re*, the second scale degree. The piece needs to resolve to the root, but Beethoven keeps us hanging there, where we least expect to be. He then runs the entire motif again, and only on the second time through does he meet our expectations. But now, that expectation is even more interesting because of the ambiguity: We wonder if, like Lucy waiting for Charlie Brown, he will pull the football of resolution away from us at the last minute.

What do we know about the neural basis for musical expectations and musical emotion? If we acknowledge that the brain is constructing a version of reality, we must reject that the brain has an accurate and strictly isomorphic representation of the world. So what is the brain holding in its neurons that represents the world around us? The brain represents all music and all other aspects of the world in terms of mental or neural codes. Neuroscientists try to decipher this code and understand its structure, and how it translates into experience. Cognitive psychologists try to understand these codes at a somewhat higher level—not at the level of neural firings, but at the level of general principles.

The way in which a picture is stored on your computer is similar, in principle, to how the neural code works. When you store a picture on your computer, the picture is not stored on your hard drive the way that a photograph is stored in your grandmother's photo album. When you open your grandmother's album, you can pick up a photo, turn it upside down, give it to a friend; it is a physical object. It is the photograph, not a representation of a photograph. On the other hand, a photo in your computer is stored in a file made up of 0s and 1s—the binary code that computers use to represent everything.

If you've ever opened a corrupt file, or if your e-mail program didn't properly download an attachment, you've probably seen a bunch of gibberish in place of what you thought was a computer file: a string of funny symbols, squiggles, and alphanumeric characters that looks like the equivalent of a comic-strip swear word. (These represent a sort of intermediate hexadecimal code that itself is resolved into 0s and 1s, but this intermediate stage is not crucial for understanding the analogy.) In the simplest case of a black-and-white photograph, a 1 might represent that there is a black dot at a particular place in the picture, and a 0 might indicate the absence of a black dot, or a white dot. You can imagine that one could easily represent a simple geometric shape using these 0s and 1s, but the 0s and 1s would not themselves be in the shape of a triangle, they would simply be part of a long line of 0s and 1s, and the computer would have a set of instructions telling it how to interpret them (and to what spatial location each number refers). If you got really good at reading such a file, you might be able to decode it, and guess what sort of image it represents. The situation is vastly more complicated with a color image, but the principle is the same. People who work with image files all the time are able to look at the stream of 0s and 1s and tell something about the nature of the photograph—not at the level of whether it is a human or a horse, perhaps, but things like how much red or gray is in the picture, how sharp the edges are, and so forth. They have learned to read the code that represents the picture.

Similarly, audio files are stored in binary format, as sequences of 0s and 1s. The 0s and 1s represent whether or not there is any sound at particular parts of the frequency spectrum. Depending on its position in the file, a certain sequence of 0s and 1s will indicate if a bass drum or a piccolo is playing.

In the cases I've just described, the computer is using a code to represent common visual and auditory objects. The objects themselves are decomposed into small components—pixels in the case of a picture, sine waves of a particular frequency and amplitude in the case of sound—and these components are translated into the code. Of course, the computer (brain) is running a lot of fancy software (mind) that translates the code effortlessly. Most of us don't have to concern ourselves with the code itself at all. We scan a photo or rip a song to our hard drive, and when we want to see it or hear it, we double-click on it and there it appears, in all its original glory. This is an illusion made possible by the many layers of translation and amalgamation going on, all of it invisible to us. This is what the neural code is like. Millions of nerves firing at different rates and different intensities, all of it invisible to us. We can't feel our nerves firing; we don't know how to speed them up, slow them down, turn them on when we're having trouble getting started on a bleary-eyed morning, or shut them off so we can sleep at night.

Years ago, my friend Perry Cook and I were astonished when we read an article about a man who could look at phonograph records and identify the piece of music that was on them, by looking at the grooves, with the label obscured. Did he memorize the patterns of thousands of record albums? Perry and I took out some old record albums and we noticed some regularities. The grooves of a vinyl record contain a code that is "read" by the needle. Low notes create wide grooves, high notes create narrow grooves, and a needle dropped inside the grooves is moving thousands of times per second to capture the landscape of the inner wall. If a person knew many pieces of music well, it would be possible to characterize them in terms of how many low notes there were (rap music has a lot, baroque concertos don't), how steady versus percussive the low notes are (think of a jazz-swing tune with walking bass as opposed to a funk tune with slapping bass), and to learn how these shapes are encoded in vinyl. This fellow's skills are extraordinary, but they're not inexplicable.

We encounter gifted auditory-code readers every day: the mechanic who can listen to the sound of your engine and determine whether your problems are due to clogged fuel injectors or a slipped timing chain; the doctor who can tell by listening to your heart whether you have an arrhythmia; the police detective who can tell when a suspect is lying by the stress in his voice; the musician who can tell a viola from a violin or a B-flat clarinet from an E-flat clarinet just by the sound. In all these cases, timbre is playing an important role in helping us to unlock the code.

How can we study neural codes and learn to interpret them? Some neuroscientists start by studying neurons and their characteristics— what causes them to fire, how rapidly they fire, what their refractory period is (how long they need to recover between firings); we study how neurons communicate with each other and the role of neurotransmitters in conveying information in the brain. Much of the work at this level of analysis concerns general principles; we don't yet know much about the neurochemistry of music, for example, although I'll reveal some exciting new results along this line from my laboratory in Chapter 5.

But I'll back up for a minute. Neurons are the primary cells of the brain; they are also found in the spinal cord and the peripheral nervous system. Activity from outside the brain can cause a neuron to fire—such as when a tone of a particular frequency excites the basilar membrane, and it in turn passes a signal up to a frequency-selective neurons in the auditory cortex. Contrary to what we thought a hundred years ago, the neurons in the brain aren't actually touching; there's a space between them called the synapse. When we say a neuron is firing, it is sending an electrical signal that causes the release of a neurotransmitter. Neurotransmitters are chemicals that travel throughout the brain and bind to receptors attached to other neurons. Receptors and neurotransmitters can be thought of as locks and keys respectively. After a neuron fires, a neurotransmitter swims across that synapse to a nearby neuron, and when it finds the lock and binds with it, that new neuron starts to fire. Not all keys fit all locks; there are certain locks (receptors) that are designed to accept only certain neurotransmitters.

Generally, neurotransmitters cause the receiving neuron to fire or prevent it from firing. The neurotransmitters are then absorbed through a process called reuptake; without reuptake, the neurotransmitters would continue to stimulate or inhibit the firing of a neuron.

Some neurotransmitters are used throughout the nervous system, and some only in certain brain regions and by certain kinds of neurons. Serotonin is produced in the brain stem and is associated with the regulation of mood and sleep. The new class of antidepressants, including Prozac and Zoloft, are known as selective serotonin reuptake inhibitors (SSRIs) because they inhibit the reuptake of serotonin in the brain, allowing whatever serotonin is already there to act for a longer period of time. The precise mechanism by which this alleviates depression, obsessive-compulsive disorder, and mood and sleep disorders is not known. Dopamine is released by the nucleus accumbens and is involved in mood regulation and the coordination of movement. It is most famous for being part of the brain's pleasure and reward system. When drug addicts get their drug of choice, or when compulsive gamblers win a bet— even when chocoholics get cocoa—this is the neurotransmitter that is released. Its role—and the important role played by the nucleus accumbens—in music was unknown until 2005.

Cognitive neuroscience has been making great leaps in understanding over the last decade. We now know so much more about how neurons work, how they communicate, how they form networks, and how neurons develop from their genetic recipes. One finding at the macro level about the function of the brain is the popular notion about hemispheric specialization—the idea that the left half of the brain and the right half of the brain perform different cognitive functions. This is certainly true, but as with much of the science that has permeated popular culture, that real story is somewhat more nuanced.

To begin with, the research on which this is based was performed on right-handed people. For reasons that aren't entirely clear, people who are left-handed (approximately 5 to 10 percent of the population) or ambidextrous sometimes have the same brain organization as right-handers, but more often have a different brain organization. When the brain organization is different, it can take the form of a simple mirror image, such that functions are simply flipped to the opposite side. In many cases, however, left-handers have a neural organization that is different in ways that are not yet well documented. Thus, any generalizations we make about hemispheric asymmetries are applicable only to the right-handed majority of the population.

Writers, businessmen, and engineers refer to themselves as left-brain dominant, and artists, dancers, and musicians as right-brain dominant. The popular conception that the left brain is analytical and the right brain is artistic has some merit, but is overly simplistic. Both sides of the brain engage in analysis and both sides in abstract thinking. All of these activities require coordination of the two hemispheres, although some of the particular functions involved are clearly lateralized.

Speech processing is primarily left-hemisphere localized, although certain global aspects of spoken language, such as intonation, emphasis, and the pitch pattern, are more often disrupted following right-hemisphere damage. The ability to distinguish a question from a statement, or sarcasm from sincerity, often rests on these right-hemisphere lateralized, nonlinguistic cues, known collectively as prosody. It is natural to wonder whether music shows the opposite asymmetry, with processing located primarily on the right. There are many cases of individuals with brain damage to the left hemisphere who lost the power of speech, but retained their musical function, and vice versa. Cases like these suggest that music and speech, although they may share some neural circuits, cannot use completely overlapping neural structures.

Local features of spoken language, such as distinguishing one speech sound from another, appear to be left-hemisphere lateralized. We've found lateralization in the brain basis of music as well. The overall contour of a melody—simply its melodic shape, while ignoring intervals—is processed in the right hemisphere, as is making fine discriminations of tones that are close together in pitch. Consistent with its language functions, the left hemisphere is involved in the naming aspects of music—such as naming a song, a performer, an instrument, or a musical interval. Musicians using their right hands or reading music from their right eye also use the left brain because the left half of the brain controls the right half of the body. There is also new evidence that tracking the ongoing development of a musical theme—thinking about key and scales and whether a piece of music makes sense or not—is lateralized to the left frontal lobes.

Musical training appears to have the effect of shifting some music processing from the right (imagistic) hemisphere to the left (logical) hemisphere, as musicians learn to talk about—and perhaps think about—music using linguistic terms. And the normal course of development seems to cause greater hemispheric specialization: Children show less lateralization of musical operations than do adults, regardless of whether they are musicians or not.

The best place to begin to look at expectation in the musical brain is in how we track chord sequences in music over time. The most important way that music differs from visual art is that it is manifested over time. As tones unfold sequentially, they lead us—our brains and our minds— to make predictions about what will come next. These predictions are the essential part of musical expectations. But how to study the brain basis of these?

Neural firings produce a small electric current, and consequently the current can be measured with suitable equipment that allows us to know when and how often neurons are firing; this is called the electroencephalogram, or EEG. Electrodes are placed (painlessly) on the surface of the scalp, much as a heart monitor might be taped to your finger, wrist, or chest. The EEG is exquisitely sensitive to the timing of neural firings, and can detect activity with a resolution of one thousandth of a second (one millisecond). But it has some limitations. EEG is not able to distinguish whether the neural activity is releasing excitatory, inhibitory, or modulatory neurotransmitters, the chemicals such as serotonin and dopamine that influence the behavior of other neurons. Because the electrical signature generated by a single neuron firing is relatively weak, the EEG only picks up the synchronous firing of large groups of neurons, rather than individual neurons.

EEG also has limited spatial resolution—that is, a limited ability to tell us the location of the neural firings, due to what is called the inverse Poisson problem. Imagine that you're standing inside a football stadium that has a large semitransparent dome covering it. You have a flashlight, and you point it up to the inside surface of the dome. Meanwhile, I'm standing on the outside, looking down at the dome from high above, and I have to predict where you're standing. You could be standing anywhere on the entire football field and shining your light at the same particular spot in the center of the dome, and from where I'm standing, it will all look the same to me. There might be slight differences in the angle or the brightness of the light, but any prediction I make about where you're standing is going to be a guess. And if you were to bounce your flashlight beam off of mirrors and other reflective surfaces before it reached the dome, I'd be even more lost. This is the case with electrical signals in the brain that can be generated from multiple sources in the brain, from the surface of the brain or deep down inside the grooves (sulci), and that can bounce off of the sulci before reaching the electrode on the outer scalp surface. Still, EEG has been helpful in understanding musical behavior because music is time based, and EEG has the best temporal resolution of the tools we commonly employ for studying the human brain.

Several experiments conducted by Stefan Koelsch, Angela Friederici, and their colleagues have taught us about the neural circuits involved in musical structure. The experimenters play chord sequences that either resolve in the standard, schematic way, or that end on unexpected chords. After the onset of the chord, electrical activity in the brain associated with musical structure is observed within 150–400 milliseconds (ms), and activity associated with musical meaning about 100–150 ms later. The structural processing—musical syntax—has been localized to the frontal lobes of both hemispheres in areas adjacent to and overlapping with those regions that process speech syntax, such as Broca's area, and shows up regardless of whether listeners have musical training. The regions involved in musical semantics—associating a tonal sequence with meaning—appear to be in the back portions of the temporal lobe on both sides, near Wernicke's area.

The brain's music system appears to operate with functional independence from the language system—the evidence comes from many case studies of patients who, postinjury, lose one or the other faculty but not both. The most famous case is perhaps that of Clive Wearing, a musician and conductor, whose brain was damaged as a result of herpes encephalitis. As reported by Oliver Sacks, Clive lost all memory except for musical memories, and the memory of his wife. Other cases have been reported for which the patient lost music but retained language and other memories. When portions of his left cortex deteriorated, the composer Ravel selectively lost his sense of pitch while retaining his sense of timbre, a deficit that inspired his writing of *Bolero*, a piece that emphasizes variations in timbre. The most parsimonious explanation is that music and language do, in fact, share some common neural resources, and yet have independent pathways as well. The close proximity of music and speech processing in the frontal and temporal lobes, and their partial overlap, suggests that those neural circuits that become recruited for music and language may start out life undifferentiated. Experience and normal development then differentiate the functions of what began as very similar neuronal populations. Consider that at a very early age, babies are thought to be synesthetic, to be unable to differentiate the input from the different senses, and to experience life and the world as a sort of psychedelic union of everything sensory. Babies may see the number five as red, taste cheddar cheeses in D-flat, and smell roses in triangles.

The process of maturation creates distinctions in the neural pathways as connections are cut or pruned. What may have started out as a neuron cluster that responded equally to sights, sound, taste, touch, and smell becomes a specialized network. So, too, may music and speech have started in us all with the same neurobiological origins, in the same regions, and using the same specific neural networks. With increasing experience and exposure, the developing infant eventually creates dedicated music pathways and dedicated language pathways. The pathways may share some common resources, as has been proposed most prominently by Ani Patel in his SSIRH—shared syntactic integration resource hypothesis.

My collaborator and friend Vinod Menon, a systems neuroscientist at Stanford Medical School, shared with me an interest in being able to pin down the findings from the Koelsch and Friederici labs, and in being able to provide solid evidence for Patel's SSIRH. For that, we had to use a different method of studying the brain, since the spatial resolution of EEG wasn't fine enough to really pinpoint the neural locus of musical syntax.

Because the hemoglobin of the blood is slightly magnetic, changes in the flow of blood can be traced with a machine that can track changes in magnetic properties. This is what a magnetic resonance imaging machine (MRI) is, a giant electromagnet that produces a report showing differences in magnetic properties, which in turn can tell us where, at any given point in time, the blood is flowing in the body. (The research on the development of the first MRI scanners was performed by the British company EMI, financed in large part from their profits on Beatles records. "I Want to Hold Your Hand" might well have been titled "I Want to Scan Your Brain.") Because neurons need oxygen to survive, and the blood carries oxygenated hemoglobin, we can trace the flow of blood in the brain too. We make the assumption that neurons that are actively firing will need more oxygen than neurons that are at rest, and so those regions of the brain that are involved in a particular cognitive task will be just those regions with the most blood flow at a given point in time. When we use the MRI machine to study the function of brain regions in this way, the technology is called functional MRI, or fMRI.

fMRI images let us see a living, functioning human brain while it is thinking. If you mentally practice your tennis serve, we can see the flow of blood move up to your motor cortex, and the spatial resolution of fMRI is good enough that we can see that it is the part of your motor cortex that controls your arm that is active. If you then start to solve a math problem, the blood moves forward, to your frontal lobes, and in particular to regions that have been identified as being associated with arithmetic problem solving, and we see this movement and ultimately the collection of blood in the frontal lobes on the fMRI scan.

Will this Frankenstein science I've just described, the science of brain imaging, ever allow us to read people's minds? I'm happy to report that the answer is probably not, and absolutely not for the foreseeable future. The reason is that thoughts are simply too complicated and involve too many different regions. With fMRI I can tell that you are listening to music as opposed to watching a silent film, but we can't yet tell if you're listening to hip-hop versus Gregorian chants, let alone what specific song you're listening to or thought you're thinking.

With the high spatial resolution of fMRI, one can tell within just a couple of millimeters where something is occurring in the brain. The problem, however, is that the temporal resolution of fMRI isn't particularly good because of the amount of time it takes for blood to become redistributed in the brain—known as hemodynamic lag. But others had already studied the *when* of musical syntax/musical structure processing; we wanted to know the *where* and in particular if the *where* involved areas already known to be dedicated to speech. We found exactly what we predicted. Listening to music and attending to its syntactic features—its structure—activated a particular region of the frontal cortex on the left side called pars orbitalis—a subsection of the region known as Brodmann Area 47. The region we found in our study had some overlap with previous studies of structure in language, but it also had some unique activations. In addition to this left hemisphere activation, we also found activation in an analogous area of the right hemisphere. This told us that attending to structure in music requires both halves of the brain, while attending to structure in language only requires the left half.

Most astonishing was that the left-hemisphere regions that we found were active in tracking musical structure were the very same ones that are active when deaf people are communicating by sign language. This suggested that what we had identified in the brain wasn't a region that simply processed whether a chord sequence was sensible, or whether a spoken sentence was sensible. We were now looking at a region that responded to sight—to the visual organization of words conveyed through American Sign Language. We found evidence for the existence of a brain region that processes structure in general, when that structure is conveyed over time. Although the inputs to this region must have come from different neural populations, and the outputs of it had to go through distinctive networks, there it was—a region that kept popping up in any task that involved organizing information over time.

The picture about neural organization for music was becoming clearer. All sound begins at the eardrum. Right away, sounds get segregated by pitch. Not much later, speech and music probably diverge into separate processing circuits. The speech circuits decompose the signal in order to identify individual phonemes—the consonants and vowels that make up our alphabet and our phonetic system. The music circuits start to decompose the signal and separately analyze pitch, timbre, contour, and rhythm. The output of the neurons performing these tasks connects to regions in the frontal lobe that put all of it together and try to figure out if there is any structure or order to the temporal patterning of it all. The frontal lobes access our hippocampus and regions in the interior of the temporal lobe and ask if there is anything in our memory banks that can help to understand this signal. Have I heard this particular pattern before? If so, when? What does it mean? Is it part of a larger sequence whose meaning is unfolding right now in front of me?

Having nailed down some of the neurobiology of musical structure and expectation, we were now ready to ask about the brain mechanisms underlying emotion and memory.

5. You Know My Name, Look Up the Number

How We Categorize Music

One of my earliest memories of music is as a three-year-old, lying on the floor underneath the family's grand piano as my mother played. Lying on our shaggy green wool carpet, with the piano above me, all I could see were my mother's legs moving the pedals up and down, but the sound—it engulfed me! It was all around, vibrating through the floor and through my body, the low notes to the right of me, the high notes to the left. The loud, dense chords of Beethoven; the flurry of dancing, acrobatic notes of Chopin; the strict, almost militaristic rhythms of Schumann, a German like my mother. In these—among my first memories of music—the sound held me in a trance, it transported me to sensory places I had never been. Time seemed to stand still while the music was playing.

How are memories of music different from other memories? Why can music trigger memories in us that otherwise seemed buried or lost? And how does expectation lead to the experience of emotion in music? How do we recognize songs we have heard before?

Tune recognition involves a number of complex neural computations interacting with memory. It requires that our brains ignore certain features while we focus only on features that are invariant from one listening to the next—and in this way, extract invariant properties of a song.

That is, the brain's computational system must be able to separate the aspects of a song that remain the same each time we hear it from those that are one-time-only variations, or from those that are peculiar to a particular presentation. If the brain didn't do this, each time we heard a song at a different volume, we'd experience it as an entirely different song! And volume isn't the only parameter that potentially changes without affecting the underlying identity of the song. Instrumentation, tempo, and pitch can be considered irrelevant from a tune-recognition standpoint. In the process of abstracting out the features that are essential to a song's identity, changes to these features must be set aside.

Tune recognition dramatically increases the complexity of the neural system necessary for processing music. Separating the invariant properties from the momentary ones is a huge computational problem. I worked for an Internet company in the late 1990s that developed software to identify MP3 files. Lots of people have soundfiles on their computers, but many of the files are either misnamed or not named at all. No one wants to go through file by file and correct bad spellings, like "Etlon John," or rename songs like "My Aim Is True" to "Alison" by Elvis Costello (the words *my aim is true* are the refrain in the chorus, but not the name of the song).

Solving this automatic naming problem was relatively easy; each song has a digital "fingerprint," and all we needed to do was to learn how to efficiently search a database of a half-million songs in order to correctly identify the song. This is called a "lookup table" by computer scientists. It is equivalent to looking up your Social Security number in a database given your name and date of birth: Only one Social Security number is presumably associated with a given name and DOB. Similarly, only one song is associated with a specific sequence of digital values that represent the overall sound of a particular performance of that song. The program works fabulously well at looking up. What it cannot do is find other versions of the same song in the database. I might have eight versions of "Mr. Sandman" on my hard drive, but if I submit the version by Chet Atkins to a program and ask it to find other versions (such as the ones by Jim Campilongo or the Chordettes), it can't. This is because the digital stream of numbers that starts the MP3 file doesn't give us anything that is readily translated to melody, rhythm, or loudness, and we don't yet know how to make this translation. Our program would have to be able to identify relative constancies in melodic and rhythmic intervals, while ignoring performance-specific details. The brain does this with ease, but no one has invented a computer that can even begin to do this.

This different ability of computers and humans is related to a debate about the nature and function of memory in humans. Recent experiments of musical memory have provided decisive clues in sorting out the true story. The big debate among memory theorists over the last hundred years has been about whether human and animal memory is relational or absolute. The relational school argues that our memory system stores information about the relations between objects and ideas, but not necessarily details about the objects themselves. This is also called the *constructivist* view, because it implies that, lacking sensory specifics, we *construct* a memory representation of reality out of these relations (with many details filled in or reconstructed on the spot). The constructivists believe that the function of memory is to ignore irrelevant details, while preserving the gist. The competing theory is called the *record-keeping* theory. Supporters of this view argue that memory is like a tape recorder or digital video camera, preserving all or most of our experiences accurately, and with near perfect fidelity.

Music plays a role in this debate because—as the Gestalt psychologists noted over one hundred years ago—melodies are defined by pitch relations (a constructivist view) and yet, they are composed of precise pitches (a record-keeping view, but only if those pitches are encoded in memory).

A great deal of evidence has accumulated in support of both viewpoints. The evidence for the constructivists comes from studies in which people listen to speech (auditory memory) or are asked to read text (visual memory) and then report what they've heard or read. In study after study, people are not very good at re-creating a word-for-word account. They remember general content, but not specific wording.

Several studies also point to the malleability of memory. Seemingly

minor interventions can powerfully affect the accuracy of memory retrieval. An important series of studies was carried out by Elizabeth Loftus of the University of Washington, who was interested in the accuracy
of witnesses' courtroom testimonies. Subjects were shown videotapes
and asked leading questions about the content. If shown two cars that
barely scraped each other, one group of subjects might be asked, "How
fast were the cars going when they scraped each other?" and another
group would be asked, "How fast were the cars going when they smashed
each other?" Such one-word substitutions caused dramatic differences
in the eyewitnesses' estimates of the speeds of the two vehicles. Then
Loftus brought the subjects back, sometimes up to a week later, and
asked, "How much broken glass did you see?" (There really was no broken glass.) The subjects who were asked the question with the word
smashed in it were more likely to report "remembering" broken glass in
the video. Their memory of what they actually saw had been reconstructed on the basis of a simple question the experimenter had asked a
week earlier.

Findings like these have led researchers to conclude that memory is
not particularly accurate, and that it is constructed out of disparate
pieces that may themselves not be accurate. Memory retrieval (and perhaps storage) undergoes a process similar to perceptual completion or
filling in. Have you ever tried to tell someone about a dream you had over
breakfast the next morning? Typically our memory of the dream appears
to us in imagistic fragments, and the transitions between elements are
not always clear. As we tell the dream, we notice gaps, and we almost
can't help but fill them in as we unfold the narrative. "I was standing on
top of a ladder outside listening to a Sibelius concert, and the sky was
raining Pez candy . . ." you might begin. But the next image is of yourself
halfway down the ladder. We naturally and automatically fill in this missing information when retelling the dream. "And I decided to protect myself from this Pez pelting, so I started climbing down the ladder where I
knew there was shelter. . . ."

This is the left brain talking (and probably the region called orbitofrontal cortex, just behind your left temple). When we fabricate a story,

it is almost always the left brain doing the fabricating. The left brain makes up stories based on the limited information it gets. Usually it gets the story right, but it will go to great lengths to sound coherent. Michael Gazzaniga discovered this in his work with commissurotomized patients—patients who had the two hemispheres of the brain surgically separated for the relief of intractable epilepsy. Much of the inputs and outputs of the brain are contralateral—the left brain controls movement in the right half of the body, and the left brain processes information that your right eye sees. A picture of a chicken's talon was shown to a patient's left brain, and a snow-covered house to his right brain (through his right and left eyes respectively). A barrier limited the sight of each eye to only one picture. The patient was then asked to select from an array of pictures the one that was most closely associated with each of the two items. The patient pointed to a chicken with his left brain (that is, his right hand) and he pointed to a shovel with his right brain. So far, so good; chicken goes with talon, and shovel with a snow-covered house. But when Gazzaniga removed the barrier and asked the patient why he had chosen the shovel, his left hemisphere saw both the chicken and the shovel and generated a story that was consistent with both images. "You need a shovel to clean out the chicken shed," the patient answered, with no awareness that he had seen a snowbound house (with his nonverbal right brain), or that he was inventing an explanation on the spot. Score another piece of evidence for the constructivists.

At MIT in the early 1960s, Benjamin White took up the mantle of the Gestalt psychologists, who wondered how it is that a song is able to retain its identity in spite of transposition in pitch and time. White systematically altered well-known songs like "Deck the Halls" and "Michael, Row Your Boat Ashore." In some cases, he would transpose all the pitches, in others he would alter the pitch distances so that contour was preserved, but the interval sizes were shrunk or stretched. He played tunes backward and forward, and changed their rhythms. In almost every case, the deformed tune was recognized more often than chance could account for.

White demonstrated that most listeners can recognize a song in transposition almost immediately and without error. And they could recognize all kinds of deformations of the original tune as well. The constructivist interpretation of this is that the memory system must be extracting some generalized, invariant information about songs and storing that. If the record-keeping account were true, they say, it would require new calculations each time we hear a song in transposition as our brains work to compare the new version to the single, stored representation we have of the actual performance. But here, it seems that memory extracts an abstract generalization for later use.

The record-keeping account follows an old idea of my favorite researchers, the Gestalt psychologists, who said that every experience leaves a trace or residue in the brain. Experiences are stored as traces, they said, that are reactivated when we retrieve the episodes from memory. A great deal of experimental evidence supports this theory. Roger Shepard showed people hundreds of photographs for a few seconds each. A week later, he brought the subjects back into the laboratory and showed them pairs of photographs that they had seen before, along with some new ones that they hadn't. In many cases, the "new" photos had only subtle differences from the old, such as the angle of the sail on a sailboat, or the size of a tree in the background. Subjects were able to remember which ones they had seen a week earlier with astonishing accuracy.

Douglas Hintzman performed a study in which people were shown letters that differed in font and capitalization. For example:

F   l   *u*   t   e

Contrary to studies of gist memory, subjects were able to remember the specific font.

We also know anecdotally that people can recognize hundreds, if not thousands, of voices. You can probably recognize the sound of your mother's voice within one word, even if she doesn't identify herself. You can tell your spouse's voice right away, and whether he or she has a cold or is angry with you, all from the timbre of the voice. Then there are well-known voices—dozens, if not hundreds, that most people can readily identify: Woody Allen, Richard Nixon, Drew Barrymore, W. C. Fields, Groucho Marx, Katharine Hepburn, Clint Eastwood, Steve Martin. We can hold in memory the sound of these voices, often as they're uttering specific content or catchphrases: "I'm not a crook," "Say the magic woid and win a hundred dollars," "Go ahead—make my day," "Well, excuuuuuse me!" We remember the specific words and specific voices, not just the gist. This supports the record-keeping theory.

On the other hand, we enjoy listening to impressionists who do comedy routines by mimicking the voices of celebrities, and often the funniest routines involve phrases that the real celebrity never said. In order for this to work, we have to have some sort of stored memory trace for the timbre of the person's voice, independent of the actual words. This could contradict the record-keeping theory by showing that it is only the abstract properties of the voice that are encoded in memory, rather than the specific details. But, we might argue that timbre is a property of sounds that is separable from other attributes; we can hold on to our "record-keeping" theory of memory by saying that we are encoding specific timbre values in memory and still explain why we can recognize the sound of a clarinet, even if it is playing a song we've never heard before.

One of the most famous cases in the neuropsychological literature is that of a Russian patient known only by his initial S, who saw the physician A. R. Luria. S. had hypermnesia, the opposite of amnesia—instead of forgetting everything, he remembered everything. S. was unable to recognize that different views of the same person were related to a single individual. If he saw a person smiling, that was one face; if the person later was frowning, that was another face. S. found it difficult to integrate the many different expressions and viewing angles of a person into a single, coherent representation of that person. He complained to Dr. Luria, "Everyone has so many faces!" S. was unable to form abstract generalizations, only his record-keeping system was intact. In order for us to understand spoken language, we need to set aside variations in how different people pronounce words, or how the same person pronounces a given phoneme as it appears in different contexts. How can the record-keeping account be consistent with this?

Scientists like having their world organized. Allowing two theories to stand that make different predictions is scientifically unappealing. We'd like to tidy up our logical world and choose one theory over the other, or generate a third, unifying theory that accounts for everything. So which account is right? Record-keeping or constructivist? In short, neither.

The research I've just described occurred contemporaneously with ground-breaking new work on categories and concepts. Categorization is a basic function of living creatures. Every object is unique, but we often act toward different objects as members of classes or categories. Aristotle laid the methods by which modern philosophers and scientists think about how concepts form in humans. He argued that categories result from lists of defining features. For example, we have in our minds an internal representation for the category "triangle." It contains a mental image or picture of every triangle we've ever seen, and we can imagine new triangles as well. At its heart, what constitutes this category and determines the boundaries of category membership (what goes in and what stays out) is a definition that might be something like this: "A triangle is a three-sided figure." If you have mathematical training, your definition might be more elaborate: "A triangle is a three-sided, closed figure, the sum of whose interior angles is 180 degrees." Subcategories of triangles might be attached to this definition, such as "an iscosceles triangle has two sides of equal length; an equilateral triangle has three sides of equal length; in a right triangle, the sum of the squares of the sides equals the square of the hypotenuse."

We have categories for all kinds of things, living and inanimate. When we're shown a new item—a new triangle, a dog we've never seen before—we assign the item to a category based on an analysis of its properties and a comparison with the category definition, according to Aristotle. From Aristotle, through to Locke and the present day, categories were assumed to be a matter of logic, and objects were either in-side or outside of a category.

After 2,300 years of no substantial work on the topic, Ludwig Wittgen-stein asked a simple question: What is a game? This launched a renais-sance of empirical work on category formation. Enter Eleanor Rosch, who did her undergraduate philosophy thesis at Reed College in Port-land, Oregon, on Wittgenstein. Rosch had planned for years to go to graduate school in philosophy, but a year with Wittgenstein, she says, "cured her" of philosophy completely. Feeling that contemporary philos-ophy had hit a dead end, Rosch wondered how she could study philo-sophical ideas empirically, how she could discover new philosophical facts. When I was teaching at UC Berkeley, where she is a professor, she told me that she thought that philosophy had done all it could do with re-spect to problems of the brain and the mind, and that experimentation was necessary to move forward. Today, following Rosch, many cognitive psychologists consider an apt description of our field to be "empirical philosophy"; that is, experimental approaches to questions and prob-lems that have been traditionally in the domain of philosophers: What is the nature of mind? Where do thoughts come from? Rosch ended up at Harvard, and took her Ph.D. there in cognitive psychology. Her doctoral thesis changed the way we think about categories.

Wittgenstein dealt the first blow to Aristotle by pulling the rug out from strict definitions of what a category is. Using the category "games" as an example, Wittgenstein argued that there is no definition or set of definitions that can encompass all games. For example, we might say that a game (a) is done for fun or recreation, (b) is a leisure activity, (c) is an activity most often found among children, (d) has certain rules, (e) is in some way competitive, (f) involves two or more people. Yet, we can generate counterexamples for each of these elements, showing that the definitions break down: (a) In the Olympic Games, are the athletes hav-ing fun? (b) Is pro football a leisure activity? (c) Poker is a game, as is jai alai, but not most often found among children. (d) A child throwing a ball against a wall is having fun, but what are the rules? (e) Ring-around-the-rosy isn't competitive. (f) Solitaire doesn't involve two or more people. How do we get out of this reliance on definitions? Is there an alternative?

Wittgenstein proposed that category membership is determined not by a definition, but by family resemblance. We call something a game if it resembles other things we have previously called games. If we go to the Wittgenstein family reunion, we might find that certain features are shared by members of the family, but that there is no single physical feature that one absolutely, positively must have to be a family member. A cousin might have Aunt Tessie's eyes; another might have the Wittgenstein chin. Some family members will have Grandpa's forehead, others will have Grandma's red hair. Rather than using a static list of definitions, family resemblance relies on a list of features that may or may not be present. The list may also be dynamic; at some point red hair may die out of the family line (if not dye out), and so we simply remove it from our list of features. If it pops up again several generations later, we can reintroduce it to our conceptual system. This prescient idea forms the basis for the most compelling theory in contemporary memory research, the multiple-trace memory models that Douglas Hintzman worked on, and which have been recently taken up by a brilliant cognitive scientist named Stephen Goldinger from Arizona.

Can we define *music* by definitions? What about types of music, such as heavy metal, classical, or country? Such attempts would certainly fail as they did for "games." We could, for example, say that heavy metal is a musical genre that has (a) distorted electric guitars; (b) heavy, loud drums; (c) three chords, or power chords; (d) sexy lead singers, usually shirtless, dripping sweat and swinging the microphone stand around the stage like it was a piece of rope; (e) ümlauts in the gröup names. But this strict list of definitions is easy to refute. Although most heavy metal songs have distorted electric guitars, so does "Beat It" by Michael Jackson—in fact, Eddie Van Halen (the heavy metal god) plays the guitar solo in that song. Even the Carpenters have a song with a distorted guitar, and no one would call them "heavy metal." Led Zeppelin—the quintessential heavy metal band and arguably the band that spawned the genre—has several songs with no distorted guitars at all ("Bron-y-aur," "Down by the Seaside," "Goin' to California," "The Battle of Nevermore"). "Stairway to Heaven" by Led Zeppelin is a heavy metal anthem, and there are no heavy, loud drums (or distorted guitars for that matter) in 90 percent of that song. Nor does "Stairway to Heaven" have only three chords. And lots of songs have three chords and power chords that are not heavy metal, including most songs by Raffi. Metallica is a heavy metal band for sure, but I've never heard anyone call their lead singer sexy, and although Mötley Crüe, Blue Öyster Cult, Motörhead, Spiñal Tap, and Queensrÿche have gratuitous umlauts, many heavy metal bands do not: Led Zeppelin, Metallica, Black Sabbath, Def Leppard, Ozzie Osbourne, Triumph, etc. Definitions of musical genres aren't very useful; we say that something is heavy metal if it resembles heavy metal—a family resemblance.

Armed with her knowledge of Wittgenstein, Rosch decided that something can be more or less a category member; rather than being all or none as Aristotle had believed, there are shades of membership, degrees of fit to a category, and subtle shadings. Is a robin a bird? Most people would answer yes. Is a chicken a bird? Is a penguin? Most people would say yes after a slight pause, but then would add that chickens and penguins are not very good examples of birds, nor typical of the category. This is reflected in everyday speech when we use linguistic hedges such as "A chicken is *technically* a bird," or "Yes, a penguin is a bird, but it doesn't fly like most other birds." Rosch, following Wittgenstein, showed that categories do not always have clear boundaries—they have fuzzy boundaries. Questions of membership are a matter of debate and there can be differences of opinion: Is white a color? Is hip-hop really music? If the surviving members of Queen perform without Freddie Mercury, am I still seeing Queen (and is it worth $150 a ticket)? Rosch showed that people can disagree about categorizations (is a cucumber a fruit or a vegetable?), and that the same person can even disagree with himself at different times about a category (is so-and-so my friend?).

Rosch's second insight was that all of the experiments on categories that had been done before her used artificial concepts and sets of artificial stimuli that had little to do with the real world. And these controlled laboratory experiments were inadvertently constructed in ways that ended up with a bias toward the experimenters' theories! This underscores an ongoing problem that plagues all of empirical science: the tension between rigorous experimental control and real-world situations. The trade-off is that in achieving one, there is often a compromise of the other. The scientific method requires that we control all possible variables in order to be able to draw firm conclusions about the phenomenon under study. Yet such control often creates stimuli or conditions that would never be encountered in the real world, situations that are so far removed from the real world as not even to be valid. The British philosopher Alan Watts, author of *The Wisdom of Insecurity*, put it this way: If you want to study a river, you don't take out a bucketful of water and stare at it on the shore. A river is not its water, and by taking the water out of the river, you lose the essential quality of river, which is its motion, its activity, its flow. Rosch felt that scientists had disrupted the flow of categories by studying them in such artificial ways. This, incidentally, is the same problem with a lot of the research that has been done in the neuroscience of music for the past decade: Too many scientists study artificial melodies using artificial sounds—things that are so removed from music, it's not clear what we're learning.

Rosch's third insight was that certain stimuli hold a privileged position in our perceptual system or our conceptual system, and that these become prototypes for a category: Categories are formed around these prototypes. In the case of our perceptual system, categories like "red" and "blue" are a consequence of our retinal physiology; certain shades of red are universally going to be regarded as more vivid, more central, than others because a specific wavelength of visible light will cause the "red" receptors in our retina to fire maximally. We form categories around these central, or focal, colors. Rosch tested this idea on a tribe of New Guinea people, the Dani, who have only two words in their language for colors, *mili* and *mola*, which essentially correspond to light and dark.

Rosch wanted to show that what we call red, and what we would pick out as an example of the best red, is not culturally determined or learned. When shown a bunch of different shades of red, we don't pick a particular one because we've been taught that it is the best red, we pick it out because our physiology bestows a privileged perceptual position on it. The Dani have no word for red in their language, and therefore no training in what constitutes a good red versus a bad red. Rosch showed her Dani subjects chips colored with dozens of different shades of red and asked them to pick out the best example of this color. They overwhelmingly selected the same "red" that Americans do, and they were better at remembering it. And they did this for other colors that they couldn't name, like greens and blues. This led Rosch to conclude that (a) categories are formed around prototypes; (b) these prototypes can have a biological or physiological foundation; (c) category membership can be thought of as a question of degree, with some tokens being "better" exemplars than others; (d) new items are judged in relation to the prototypes, forming gradients of category membership; and the final blow for Aristotelian theory, (e) there don't need to be any attributes which all category members have in common, and boundaries don't have to be definite.

We've done some informal experiments in my laboratory with musical genres and have found similar results. People appear to agree as to what are prototypical songs for musical categories, such as "country music," "skate punk," and "baroque music." They are also inclined to consider certain songs or groups as less good examples than the prototype: the Carpenters aren't *really* rock music; Frank Sinatra is not *really* jazz, or at least not as much as John Coltrane is. Even within the category of a single artist, people make graded distinctions that imply a prototype structure. If you ask me to pick out a Beatles song, and I select "Revolution 9" (an experimental tape piece written by John Lennon and Paul McCartney, with no original music, no melody or rhythm, which begins with an announcer repeating, "Number 9, Number 9," over and over again) you might complain that I was being difficult. "Well, *technically* that's a

Beatles piece—but that's not what I meant!" Similarly, Neil Young's one album of fifties doo-wop (*Everybody's Rockin'*) is not representative (or typical) Neil Young; Joni Mitchell's jazz foray with Charles Mingus is not what we usually think of when we think of Joni Mitchell. (In fact, Neil Young and Joni Mitchell were each threatened with contract cancellations by their record labels for making music that was not deemed Neil Young–like and Joni Mitchell–like, respectively.)

Our comprehension of the world around us begins with specific and individual cases—a person, a tree, a song—and through experience with the world, these particular objects are almost always dealt with in our brains as members of a category. Roger Shepard has described the general issue in all of this discussion in terms of evolution. There are three basic appearance-reality problems that need to be solved by all higher animals, he says. In order to survive, to find edible food, water, shelter, to escape predators, and to mate, the organism must deal with three scenarios.

First, objects, though in presentation they may be similar, are inherently different. Objects that may create identical, or nearly identical, patterns of stimulation on our eardrums, retinas, taste buds, or touch sensors may actually be different entities. The apple I saw on the tree is different from the one I am holding in my hand. The different violin sounds I hear coming from the symphony, even when they're all playing the same note, represent several different instruments.

Second, objects, though in presentation they may be different, are inherently identical. When we look at an apple from above, or from the side, it appears to be an entirely different object. Successful cognition requires a computational system that can integrate these separate views into a coherent representation of a single object. Even when our sensory receptors receive distinct and nonoverlapping patterns of activation, we need to abstract out information that is critical to creating a unified representation of the object. Although I may be used to hearing your voice in person, through both ears, when I hear you over the phone, in one ear, I need to recognize that you're the same person.

The third appearance-reality problem invokes higher-order cognitive processes. The first two are perceptual processes: understanding that a single object may manifest itself in multiple viewpoints, or that several objects may have (nearly) identical viewpoints. The third problem states that objects, although different in presentation, are of the same natural kind. This is an issue in categorization, and it is the most powerful and advanced principle of all. All higher mammals, many lower mammals and birds, and even fish, can categorize. Categorization entails treating objects that appear different as of the same kind. A red apple may look different from a green apple, but they are both still apples. My mother and father may look very different, but they are both caregivers, to be trusted in an emergency.

Adaptive behavior, then, depends on a computational system that can analyze the information available at the sensory surfaces into (1) the invariant properties of the external object or scene, and (2) the momentary circumstances of the manifestation of that object or scene. Leonard Meyer notes that classification is essential to enable composers, performers, and listeners to internalize the norms governing musical relationships, and consequently, to comprehend the implications of patterns, and experience deviations from stylistic norms. Our need to classify, as Shakespeare says in *A Midsummer Night's Dream*, is to give "to airy nothing/A local habitation and a name."

Shepard's characterization recast the categorization problem as an evolutionary/adaptive one. In the meantime, Rosch's work was beginning to shake up the research community, and dozens of leading cognitive psychologists began to study to challenge her theory. Posner and Keele had shown that people store prototypes in memory. In a clever experiment, they created tokens that contained patterns of dots placed in a square— something like the face of dice, but with the dots more or less randomly placed on each face. They called these the prototypes. Then they shifted some of the dots a millimeter or so in one random direction or another. This created a set of distortions from the prototype—that is, variations—

that differed in their relationship to the prototype. Due to random variation, some of the tokens could not be easily identified with one prototype or another, the distortions were just too great.

This is like what a jazz artist does with a well-known song, or standard. When we compare Frank Sinatra's version of "A Foggy Day" with the version by Ella Fitzgerald and Louis Armstrong, we hear that some of the pitches and rhythms are the same and some are different; we expect a good vocalist to interpret the melody, even if that means changing it from the way the composer originally wrote it. In the courts of Europe during the baroque and enlightenment eras, musicians like Bach and Haydn would regularly perform variations of themes. Aretha Franklin's version of "Respect" differs from that written and performed by Otis Redding in interesting ways—but we still consider it the same song. What does this say about prototypes and the nature of categories? Can we say that the musical variations share a family resemblance? Are each of these versions of a song variations on an ideal prototype?

Posner and Keele addressed the general question of categories and prototypes using their dot stimuli. Subjects were shown pieces of paper with version after version of these squares with dots in them, each of them different, but they were never shown the prototypes from which the variations were made. The subjects weren't told how these dots patterns had been constructed, or that prototypes for these various forms existed. A week later, they asked the subjects to look at more pieces of paper, some old and some new, and to indicate which ones they had seen before. The subjects were good at identifying which ones they had seen before and which ones they hadn't. Now, unbeknownst to the subjects, Posner and Keele had slipped in the prototypes from which all the figures had been derived. Astonishingly, the subjects often identified the two previously unseen prototypes as figures they had seen before. This provided the foundation for an argument that prototypes are stored in memory—how else could the subjects have misidentified the unseen tokens? In order to store in memory something that wasn't seen, the memory system must be performing some operations on the stimuli; there must be a form of processing going on at some stage that goes beyond merely preserving the information that was presented. This seemed like the death of any record-keeping theory; if prototypes are stored in memory, memory must be constructive.

What we learned from Ben White, and subsequent work by Jay Dowling of the University of Texas and others, is that music is quite robust in the face of transformations and distortions of its basic features. We can change all of the pitches used in the song (transposition), the tempo, and the instrumentation, and the song is still recognized as the same song. We can change the intervals, the scales, even the tonality from major to minor or vice versa. We can change the arrangement—say from bluegrass to rock, or heavy metal to classical—and, as the Led Zepplin lyric goes, the song remains the same. I have a recording of a bluegrass group, the Austin Lounge Lizards, playing "Dark Side of the Moon" by the progressive rock group Pink Floyd, using banjos and mandolins. I have recordings of the London Symphony Orchestra playing the songs of the Rolling Stones and Yes. With such dramatic changes, the song is still recognizable as the song. It seems, then, that our memory system extracts out some formula or computational description that allows us to recognize songs in spite of these transformations. It seems that the constructivist account most closely fits the music data, and from Posner and Keele, it fits visual cognition as well.

In 1990, I took a course at Stanford called "Psychoacoustics and Cognitive Psychology for Musicians," jointly offered by the departments of music and psychology. The course was team-taught by an all-star cast: John Chowning, Max Mathews, John Pierce, Roger Shepard, and Perry Cook. Each student had to complete a research project, and Perry suggested that I look at how well people can remember pitches, and specifically whether they can attach arbitrary labels to those pitches. This experiment would unite memory and categorization. The prevailing theories predicted that there was no reason for people to retain absolute pitch information—the fact that people can so easily recognize tunes in transposition argues for that. And most people cannot name the notes, except for the one in ten thousand who have absolute pitch.

Why is absolute pitch (AP) is so rare? People with AP can name notes as effortlessly as most of us name colors. If you play someone with AP a C-sharp on the piano, he or she can tell you it was a C-sharp. Most people can't do that, of course—even most musicians can't unless they're looking at your fingers. Most AP possessors can name the pitch of other sounds, too, like car horns, the hum of fluorescent lights, and knives clinking against dinner plates. As we saw earlier, color is a psychophysical fiction—it doesn't exist in the world, but our brains impose a categorical structure, such as broad swatches of red or blue, on the unidimensional continuum of frequency of light waves. Pitch is also a psychophysical fiction, the consequence of our brains' imposing a structure on the unidimensional continuum of frequency of the sound waves. We can instantly name a color just by looking at it. Why can't we name sounds just by listening to them?

Well, most of us can identify sounds as effortlessly as we identify colors; it's simply not the pitch we identify, but rather, the timbre. We can instantly say of a sound, "That's a car horn," or "That's my grandmother Sadie with a cold," or "That's a trumpet." We can identify tonal color, just not pitch. Still, it remains an unsolved problem why some people have AP and others don't. The late Dixon Ward from the University of Minnesota noted wryly that the real question isn't "Why do only a few people have AP?" but "Why don't we all?"

I read everything I could about AP. In the 130 years from 1860 to 1990, roughly a hundred research articles were published on the subject. In the fifteen years since 1990 there has been an equal number! I noticed that all the AP tests required the subjects to use a specialized vocabulary—the note names—that only musicians would know. There seemed to be no way to test for absolute pitch among nonmusicians. Or was there?

Perry suggested that we find out how easily the proverbial man in the street could learn to name pitches by associating particular pitches with arbitrary names, like Fred or Ethel. We thought about using piano notes, pitch pipes, and all kinds of things (except for kazoos, for obvious reasons), and decided that we'd get a bunch of tuning forks and hand them out to nonmusicians. Subjects were instructed to bang the tuning forks against their knees several times a day for a week, hold it up to their ears, and try to memorize the sound. We told half the people that their sound was called Fred and we told the other half it was called Ethel (after the neighbors of Lucy and Ricky on *I Love Lucy*; their last name was Mertz, which rhymes with Hertz, a pleasing coincidence that we didn't realize until years later).

Half of each group had forks tuned to middle C, the other half had forks tuned to G. We turned them loose, then took the forks away from them for a week, and then had them come back into the laboratory. Half of the subjects were asked to sing back "their pitch" and half were asked to pick it out from three notes that I played on a keyboard. The subjects were overwhelmingly able to reproduce or recognize "their" note. This suggested to us that ordinary people could remember notes with arbitrary names.

This got us thinking about the role that names play in memory. Although the course was over and I had handed in my term paper, we were still curious about this phenomenon. Roger Shepard asked if nonmusicians might be able to remember the pitches of songs even though they don't have names for them. I told him about a study by Andrea Halpern. Halpern had asked nonmusicians to sing well-known songs such as "Happy Birthday" or "Frère Jacques" from memory on two different occasions. She found that although people tended not to sing in the same keys as one another, they did tend to sing a song consistently, in the same key from one occasion to the other. This suggested that they had encoded the pitches of the songs in long-term memory.

Naysayers suggested that these results could be accounted for without memory for pitch if the subjects had simply relied on muscle memory for the position of their vocal chords from one time to another. (To me, muscle memory is still a form of memory—labeling the phenomenon does nothing to change it.) But an earlier study by Ward and his colleague Ed Burns from the University of Washington had shown that muscle memory isn't actually all that good. They asked trained singers with absolute pitch to "sight-read" from a musical score; that is, the singers had to look at music they had never seen before and sing it using their knowledge of absolute pitch and their ability to read music. This is something they're usually very good at. Professional singers can sight-sing if you give them a starting pitch. Only professional singers with AP, however, can sing in the right key just by looking at the score; this is because they have some internal template, or memory, for how note names and sounds match up with each other—that's what AP is. Now, Ward and Burns had their AP singers wear headphones, and they blasted the singers with noise so that they couldn't hear what they were singing—they had to rely on muscle memory alone. The surprising finding was that their muscle memory didn't do very well. On average, it only got them to within a third of an octave of the correct tone.

We knew that nonmusicians tended to sing consistently. But we wanted to push the notion further—how accurate is the average person's memory for music? Halpern had chosen well-known songs that don't have a "correct" key—each time we sing "Happy Birthday," we're likely to sing it in a different key; someone begins on whatever pitch first comes to mind and we follow. Folk and holiday songs are sung so often and by so many people that they don't have an objectively correct key. This is reflected in the fact that there is no standard recording that could be thought of as a reference for these songs. In the jargon of my field, we would say that a single canonical version does not exist.

The opposite is true with rock/pop songs. Songs by the Rolling Stones, the Police, the Eagles, and Billy Joel do exist in a single canonical version. There is one standard recording (in most cases) and that is the only version anyone has ever heard (with the exception of the occasional bar band playing the song, or if we go see the group live). We've probably heard these songs as many times as we've heard "Deck the Halls." But every time we've heard, say, M. C. Hammer's "U Can't Touch This" or U2's "New Year's Day," they've been in the same key. It is difficult to recall a version other than the canonical one. After hearing a song thousands of times, might the actual pitches become encoded in memory?

To study this, I used Halpern's method of simply asking people to sing their favorite songs. I knew from Ward and Burns that their muscle memory wouldn't be good enough to get them there. In order to reproduce the correct key, they'd have to be keeping stable, accurate memory traces of pitches in their heads. I recruited forty nonmusicians from around campus and asked them to come into the laboratory and sing their favorite song from memory. I excluded songs that existed in multiple versions and songs that had been recorded more than once, which would exist out-there-in-the-world in more than one key. I was left with songs for which there is a single well-known recording that is the standard, or reference—songs such as "Time and Tide" by Basia or "Opposites Attract" by Paula Abdul (this was 1990, after all), as well as songs such as "Like a Virgin" by Madonna and "New York State of Mind" by Billy Joel.

I recruited subjects with a vague announcement for a "memory experiment." Subjects would receive five dollars for ten minutes. (This is usually how cognitive psychologists get subjects, by putting up notices around campus. We pay more for brain imaging studies, usually around fifty dollars, just because it is somewhat unpleasant to be in a confined, noisy scanner.) A lot of subjects complained vociferously upon discovering the details of the experiment. They weren't singers, they couldn't carry a tune in a bucket, they were afraid they'd ruin my experiment. I persuaded them to try anyway. The results were surprising. The subjects tended to sing at, or very near, the absolute pitches of their chosen songs. I asked them to sing a second song and they did it again.

This was convincing evidence that people were storing absolute pitch information in memory; that their memory representation did not just contain an abstract generalization of the song, but details of a particular performance. In addition to singing with the correct pitches, other performance nuances crept in; subjects' reproductions were rich with the vocal affectations of the original singers. For example, they would reproduce the high-pitched "ee-ee" of Michael Jackson in "Billie Jean," or the enthusiastic "Hey!" of Madonna in "Like a Virgin"; the syncopation of Karen Carpenter in "Top of the World" as well as the raspy voice of Bruce Springsteen on the first word of "Born in the U.S.A." I created a tape that had the subjects' productions on one channel of a stereo signal and the original recording on the other; it sounded as though the subjects were singing along with the record—but we hadn't played the record to them, they were singing along with the memory representation in their head, and that memory representation was astonishingly accurate.

Perry and I also found that the majority of subjects sang at the correct tempo. We checked to see if all the songs were merely sung at the same tempo to begin with, which would mean that people had simply encoded in memory a single, popular tempo. But this wasn't the case, there was a large range of tempos. In addition, in their own subjective accounts of the experiment, the subjects told us that they were "singing along with an image" or "recording" inside their heads. How does this mesh with a neural account of the findings?

By now I was in graduate school with Mike Posner and Doug Hintzman. Posner, always on the watch for neural plausibility, told me about the newest work of Petr Janata. Petr had just completed a study in which he kept track of people's brain waves while they listened to music and while they imagined music. He used EEG, placing sensors that measure electrical activity emanating from the brain across the surface of the scalp. Both Petr and I were surprised to see that it was nearly impossible to tell from the data whether people were listening to or imagining music. The pattern of brain activity was virtually indistinguishable. This suggested that people use the same brain regions for remembering as they do for perceiving.

What does this mean exactly? When we perceive something, a particular pattern of neurons fire in a particular way for a particular stimulus. Although smelling a rose and smelling rotten eggs both invoke the olfactory system, they use different neural circuits. Remember, neurons can connect to one another in millions of different ways. One configuration of a group of olfactory neurons may signal "rose" and another may signal "rotten eggs." To add to the complexity of the system, even the same neurons may have different settings associated with a different event-in-the-world. The act of perceiving then entails that an interconnected set of neurons becomes activated in a particular way, giving rise to our mental representation of the object that is out-there-in-the-world. Remembering may simply be the process of recruiting that same group of neurons we used during perception to help us form a mental image during recollection. We re-member the neurons, pulling them together again from their disparate locations to become members of the original club of neurons that were active during perception.

The common neural mechanisms that underlie perception of music and memory for music help to explain how it is that songs get stuck in our heads. Scientists call these *ear worms*, from the German *Ohrwurm*, or simply the stuck song syndrome. There has been relatively little scientific work done on the topic. We know that musicians are more likely to have ear worm attacks than nonmusicians, and that people with obsessive-compulsive disorder (OCD) are more likely to report being troubled by ear worms—in some cases medications for OCD can minimize the effects. Our best explanation is that the neural circuits representing a song get stuck in "playback mode," and the song—or worse, a little fragment of it—plays back over and over again. Surveys have revealed that it is rarely an entire song that gets stuck, but rather a piece of the song that is typically less than or equal in duration to the capacity of auditory short-term ("echoic") memory: about 15 to 30 seconds. Simple songs and commercial jingles seem to get stuck more often than complex pieces of music. This predilection for simplicity has a counterpart in our formation of musical preference, which I'll discuss in Chapter 8.

The findings from my study of people singing their favorite songs with accurate pitch and tempo have been replicated by other laboratories, so we know now that they're not just the result of chance. Glenn Schellenberg at the University of Toronto—incidentally, an original member of the New Wave group Martha and the Muffins—performed an extension of my study in which he played people snippets of Top 40 songs that lasted a tenth of a second or so, about the same duration as a finger snap. People were given a list of song names and had to match them up with the snippet they heard. With such a short excerpt, they could not rely on the melody or rhythm to identify the songs—in every case, the excerpt was less than one or two notes. The subjects could only rely on timbre, the overall sound of the song. In the introduction, I mentioned the importance that timbre holds for composers, songwriters, and producers.

Paul Simon thinks in terms of timbre; it is the first thing he listens for in his music and the music of others. Timbre also appears to hold this privileged position for the rest of us; the nonmusicians in Schellenberg's study were able to identify songs using only timbral cues a significant percentage of the time. Even when the excerpts were presented backward, so that anything overtly familiar was disrupted, they still recognized the songs.

If you think about the songs that you know and love, this should hold some intuitive sense. Quite apart from the melody, the specific pitches and rhythms, some songs simply have an overall sound, a sonic color. It is similar to that quality that makes the plains of Kansas and Nebraska look one way, the coastal forests of northern California, Oregon, and Washington another, the mountains of Colorado and Utah yet another. Before recognizing any details in a picture of these places, you apprehend the overall scene, the landscape, the way that things look together. The auditory landscape, the soundscape, also has a presentation that is unique in much of the music we hear. Sometimes it is not song specific. This is what allows us to identify musical groups even when we cannot recognize a specific song. Early Beatles albums have a particular timbral quality such that many people can identify a recording as the Beatles if they don't immediately recognize the song—even if it is a song they never heard before. This same quality allows us to identify imitations of the Beatles, when Eric Idle and his colleagues from Monty Python formed the fictitious group the Rutles as a Beatles satire band, for example. By incorporating many of the distinctive timbral elements of the Beatles soundscape, they were able to create a realistic satire that sounds like the Beatles.

Overall timbral presentations, soundscapes, can also apply to whole eras of music. Classical records from the 1930s and early 1940s have a particular sound to them due to the recording technology of the day. Nineteen eighties rock, heavy metal, 1940s dance hall music, and late 1950s rock and roll are fairly homogeneous eras or genres. Record producers can re-create these sounds in the studio by paying close attention to details of the soundscape: the microphones they use, the way they mix instruments, and so on. And many of us can hear a song and accurately guess what era it belongs to. One clue is often the echo, or reverberation, used on the voice. Elvis Presley and Gene Vincent had a very distinctive "slap-back" echo, in which you hear a sort of instant repeat of the syllable the vocalist just sang. You hear it on "Be-Bop-A-Lula" by Gene Vincent and by Ricky Nelson, on "Heartbreak Hotel" by Elvis, and on "Instant Karma" by John Lennon. Then there is the rich, warm echo made by a large tiled room on recordings by the Everly Brothers, such as "Cathy's Clown" and "Wake Up Little Susie." There are many distinctive elements in the overall timbre of these records that we identify with the era in which they were made.

Taken together, the findings from memory for popular songs provide strong evidence that absolute features of music are encoded in memory. And there is no reason to think that musical memory functions differently from, say, visual, olfactory, tactile, or gustatory memory. It would seem, then, that the record-keeping hypothesis has enough support for us to adopt it as a model for how memory works. But before we do, what do we do with the evidence supporting the constructivist theory? Since people can so readily recognize songs in transposition, we need to account for how this information is stored and abstracted. And there is yet another feature of music that is familiar to all of us, which an adequate theory of memory needs to account for: We can scan songs in our mind's ear and we can imagine transformations of them.

Here's a demonstration, based on an experiment that Andrea Halpern conducted: Does the word *at* appear in the American national anthem ("The Star-Spangled Banner")? Think about it before you read on.

If you're like most people, you "scanned" through the song in your head, singing it to yourself at a rapid rate, until you got to the phrase "What so proudly we hailed, *at* the twilight's last gleaming." Now, a number of interesting things happened here. First, you probably sang the song to yourself faster than you've ever heard it. If you were only able to play back a particular version you had stored in memory, you wouldn't be able to do this. Second, your memory is not like a tape recorder; if you want to speed up a tape recorder or video or film to make the song go faster, you have to also raise the pitch. But in our minds, we can vary pitch and tempo independently. Third, when you did finally reach the word *at* in your mind—your "target" in answering the question I posed— you probably couldn't help yourself from continuing, pulling up the rest of the phrase, "the twilight's last gleaming." This suggests that our memory for music involves hierarchical encoding—not all words are equally salient, and not all parts of a musical phrase hold equal status. We have certain entry points and exit points that correspond to specific phrases in the music—again, unlike a tape recorder.

Experiments with musicians have confirmed this notion of hierarchical encoding in other ways. Most musicians cannot start playing a piece of music they know at any arbitrary location; musicians learn music according to a hierarchical phrase structure. Groups of notes form units of practice, these smaller units are combined into larger units, and ultimately into phrases; phrases are combined into structures such as verses and choruses or movements, and ultimately everything is strung together as a musical piece. Ask a performer to begin playing from a few notes before or after a natural phrase boundary, and she usually cannot do it, even when reading from a score. Other experiments have shown that musicians are faster and more accurate at recalling whether a certain note appears in a musical piece if that note is at the beginning of a phrase or is on a downbeat, rather than being in the middle of a phrase or on a weak beat. Even musical notes appear to fall into categories, as to whether they are the "important" notes of a piece or not. Many amateur singers don't store in memory every note of a musical piece. Rather, we store the "important" tones—even without any musical training, we all have an accurate and intuitive sense of which those are—and we store musical contour. Then, when it comes time to sing, the amateur knows that she needs to go from this tone to that tone, and she fills in the missing tones on the spot, without having explicitly memorized each of them. This reduces memory load substantially, and makes for greater efficiency.

From all of these phenomena, we can see that a principal development in memory theory over the last hundred years was its convergence with the research on concepts and categories. One thing is for sure now: Our decision about which memory theory is right—the constructivist or the record-keeping/tape-recorder theory—will have implications for theories of categorization. When we hear a new version of our favorite song, we recognize that it is fundamentally the same song, albeit in a different presentation; our brains place the new version in a category whose members include all the versions of that song we've heard.

If we're real music fans, we might even displace a prototype in favor of another based on knowledge that we gain. Take, for example, the song "Twist and Shout." You might have heard it countless times by live bands in various bars and Holiday Inns, and you might also have heard the recordings by the Beatles and the Mamas and the Papas. One of these latter two versions may even be your prototype for the song. But if I tell you that the Isley Brothers had a hit with the song two years before the Beatles recorded it, you might reorganize your category to accommodate this new information. That you can accomplish such reorganization based on a top-down process suggests that there is more to categories than Rosch's prototype theory states. Prototype theory has a close connection to the constructivist theory of memory, in that details of individual cases are discarded, and the gist or abstract generalization is stored—both in the sense of what is being stored as a memory trace, and what is being stored as the central memory of the category.

The record-keeping memory account has a correlate in categorization theory, too, and it is called exemplar theory. As important as prototype theory was, and as well as it accounted for both our intuitions and experimental data on category formation, scientists started to find problems with it in the 1980s. Led by Edward Smith, Douglas Medin, and Brian Ross, researchers identified some weaknesses in prototype theory. First, when the category is broad and category members differ widely, how can there be a prototype? Think, for example, of the category "tool." What is the prototype for it? Or for the category "furniture"? What is the prototypical song by a female pop artist?

Smith, Medin, Ross, and their colleagues also noticed that within these kinds of heterogeneous categories, context can have a strong impact on what we consider to be the prototype. The prototypical tool at an automobile repair garage is more likely to be a wrench than a hammer, but at a home construction site the opposite would be true. What is the prototypical instrument in a symphony orchestra? I'm willing to bet that you didn't say "guitar" or "harmonica," but asked the same question for a campfire I doubt you would say "French horn" or "violin."

Contextual information is part of our knowledge about categories and category members, and prototype theory doesn't account for this. We know, for example, that within the category "birds" the ones that sing tend to be small. Within the category "my friends," there are some that I would let drive my car and some I wouldn't (based on their accident history and whether or not they have a license). Within the category "Fleetwood Mac songs," some are sung by Christine McVie, some by Lindsey Buckingham, and some by Stevie Nicks. Then there is knowledge about the three distinct eras of Fleetwood Mac: the blues years with Peter Green on guitar, the middle pop years with Danny Kirwan, Christine McVie, and Bob Welch as songwriters, and the later years after Buckingham-Nicks joined. If I ask you for the prototypical Fleetwood Mac song, context is important. If I ask you for the prototypical Fleetwood Mac member, you'll throw up your hands and tell me there is something wrong with the question! Although Mick Fleetwood and John McVie, the drummer and bassist, are the only two members who have been with the group from its beginning, it doesn't seem quite right to say that the prototypical member of Fleetwood Mac is the drummer or the bassist, neither of whom sings or wrote the major songs. Contrast this with the Police, for whom we might say that Sting was the prototypical member, as songwriter, singer, and bassist. But if someone said that, you could just as forcefully argue that she's wrong, Sting is not the prototypical member, he is merely the best known and the most crucial member, not the same thing. The trio we know as the Police is a small but heterogeneous category, and to talk about a prototypical member doesn't seem to be in keeping with the spirit of what a prototype is—the central tendency, the average, the seen or unseen object that is most typical of the category. Sting is not typical of the Police in the sense of being any kind of average;

he is rather atypical in that he is so much better known than the other two, Andy Summers and Stewart Copeland, and his history since the Police has followed such a different course.

Another problem is that, although Rosch doesn't explicitly state this, her categories seem to take some time to form. Although she explicitly allows for fuzzy boundaries, and the possibility that a given object could occupy more than one category ("chicken" could occupy the categories "bird," "poultry," "barnyard animals," and "things to eat"), there isn't a clear provision for our being able to make up new categories on the spot. And we do this all the time. The most obvious example is when we make playlists for our MP3 players, or load up our car with CDs to listen to on a long drive. The category "music I feel like listening to now" is certainly a new and dynamic one. Or consider this: What do the following items have in common: children, wallet, my dog, family photographs, and car keys? To many people, these are *things to take with me in the event of a fire.* Such collections of things form ad hoc categories, and we are adept at making these. We form them not from perceptual experience with things-in-the-world, but from conceptual exercises such as the ones above.

I could form another ad hoc category with the following story: "Carol was in trouble. She had spent all her money and she wouldn't be getting a paycheck for another three days. There was no food in the house." This leads to the ad hoc functional category "ways to get food for the next three days," which might include "go to a friend's house," "write a bad check," "borrow money from someone," or "sell my copy of *This Is Your Brain on Music.*" Thus, categories are formed not just by matching properties, but by theories about how things are related. We need a theory of category formation that will account for (a) categories that have no clear prototype, (b) contextual information, and (c) the fact that we form new categories all the time, on the spot. To accomplish this, it seems that we must have retained some of the original information from the items, because you never know when you're going to need it. If (according to the constructivists) I'm only storing abstract, generalized gist information, how could I construct a category like "songs that have the word *love* in them without having the word *love* in the title"? For example, "Here,

There and Everywhere" (the Beatles), "Don't Fear the Reaper" (Blue Öyster Cult), "Something Stupid" (Frank and Nancy Sinatra), "Cheek to Cheek" (Ella Fitzgerald and Louis Armstrong), "Hello Trouble (Come On In)" (Buck Owens), "Can't You Hear Me Callin'" (Ricky Skaggs).

Prototype theory suggests the constructivist view, that an abstract generalization of the stimuli we encounter becomes stored. Smith and Medin proposed exemplar theory as an alternative. The distinguishing feature of exemplar theory is that every experience, every word heard, every kiss shared, every object seen, every song you've ever listened to, is encoded as a trace in memory. This is the intellectual descendant of the so-called residue theory of memory proposed by the Gestalt psychologists.

Exemplar theory accounts for how we are able to retain so many details with such accuracy. Under it, details and context are retained in the conceptual memory system. Something is judged a member of a category if it resembles other members of that category more than it resembles members of an alternative, competing category. Indirectly, exemplar theory can also account for the experiments that suggested that prototypes are stored in memory. We decide whether a token is a member of a category by comparing it with all the other category members—memories of everything we encountered that is a category member and every time we encountered it. If we are presented with a previously unseen prototype—as in the Posner and Keele experiment—we categorize it correctly and swiftly because it bears a maximum resemblance to all the other stored examples. The prototype will be similar to examples from its own category and not similar to examples from alternative categories, so it reminds you of examples from the correct category. It makes more matches than any previously seen example because, by definition, the prototype is the central tendency, the average category member. This has powerful implications for how we come to enjoy new music we've never heard before, and how we can like a new song instantly—the topic of Chapter 6.

The convergence of exemplar theory and memory theory comes in the form of a relatively new group of theories, collectively called "multiple-trace memory models." In this class of models, each experience we have is preserved with high fidelity in our long-term memory system. Memory distortions and confabulations occur when, in the process of retrieving a memory, we either run into interference from other traces that are competing for our attention—traces with slightly different details—or some of the details of the original memory trace have degraded due to normally occurring neurobiological processes.

The true test of such models is whether they can account for and predict the data on prototypes, constructive memory, and the formation and retention of abstract information—such as when we recognize a song in transposition. We can test the neural plausibility of these models through neuroimaging studies. The director of the U.S. NIH (National Institutes of Health) brain laboratories, Leslie Ungerleider, and her colleagues performed fMRI studies showing that representations of categories are located in specific parts of the brain. Faces, animals, vehicles, foods, and so on have been shown to occupy specific regions of the cortex. And based on lesion studies, we've found patients who have lost the ability to name members of some categories, while other categories remain intact. These data speak to the reality of conceptual structure and conceptual memory in the brain; but what about the ability to store detailed information and still end up with a neural system that acts like it has stored abstractions?

In cognitive science, when neurophysiological data is lacking, neural network models are often used to test theories. These are essentially brain simulations that run on computers, with models of neurons, neuronal connections, and neuronal firings. The models replicate the parallel nature of the brain, and so are often referred to as parallel distributed processing or PDP models. David Rumelhardt from Stanford and Jay McClelland from Carnegie Mellon University were at the forefront of this type of research. These aren't ordinary computer programs. PDP models operate in parallel (like real brains), they have several layers of processing units (as do the layers of the cortex), the simulated neurons can be connected in myriad different ways (like real neurons), and simulated neurons can be pruned out of the network or added into the network as necessary (just as the brain reconfigures neural networks as incoming information arrives). By giving PDP models problems to solve—such as categorization or memory storage and retrieval problems—we can learn whether the theory in question is plausible; if the PDP model acts the way humans do, we take that as evidence that things may work in humans that way as well.

Douglas Hintzman built the most influential PDP model demonstrating the neural plausibility of multiple-trace memory models. His model, named MINERVA after the Roman goddess of knowledge, was introduced in 1986. She stored individual examples of the stimuli she encountered, and still managed to produce the kind of behavior we would expect to see from a system that stored only prototypes and abstract generalizations. She did this in much the way that Smith and Medin describe, by comparing new instances to stored instances. Stephen Goldinger found further evidence that multiple-trace models can produce abstractions with auditory stimuli, specifically with words spoken in specific voices.

There is now an emerging consensus among memory researchers that neither the record-keeping nor the constructivist view is correct, but that a third view, a hybrid of sorts, is the correct theory: the multiple-trace memory model. The experiments on the accuracy of memory for musical attributes are consistent with the Hintzman/Goldinger multiple-trace models. This is the model that most closely resembles the exemplar model of categorization, for which there is also an emerging consensus.

How does a multiple-trace memory model account for the fact that we extract invariant properties of melodies as we are listening to them? As we attend to a melody, we must be performing calculations on it; in addition to registering the absolute values, the details of its presentation—details such as pitch, rhythms, tempo, and timbre—we must also be calculating melodic intervals and tempo-free rhythmic information. Neuroimaging studies from Robert Zatorre and his colleagues at McGill have suggested this is the case. Melodic "calculation centers" in the dorsal (upper) temporal lobes—just above your ears—appear to be paying attention to interval size and distances between pitches as we listen to music, creating a pitch-free template of the very melodic values we will need in order to recognize songs in transposition. My own neuroimaging studies have shown that familiar music activates both these regions and the hippocampus, a structure deep in the center of the brain that is known to be crucial to memory encoding and retrieval. Together, these findings suggest that we are storing both the abstract and the specific information contained in melodies. This may be the case for all kinds of sensory stimuli.

Because they preserve context, multiple-trace memory models can also explain how we sometimes retrieve old and nearly forgotten memories. Have you ever been walking down the street and suddenly smelled an odor that you hadn't smelled in a long time, and that triggered a memory of some long-ago event? Or heard an old song come on the radio that instantly retrieved deeply buried memories associated with when that song was first popular? These phenomena get to the heart of what it means to have memories. Most of us have a set of memories that we treat something like a photo album or scrapbook. Certain stories we are accustomed to telling to our friends and families, certain past experiences we recall for ourselves during times of struggle, sadness, joy, or stress, to remind us of who we are and where we've been. We can think of this as the repertoire of our memories, those memories that we are used to playing back, something like the repertoire of a musician and the pieces he knows how to play.

According to the multiple-trace memory models, every experience is potentially encoded in memory. Not in a particular place in the brain, because the brain is not like a warehouse; rather, memories are encoded in groups of neurons that, when set to proper values and configured in a particular way, will cause a memory to be retrieved and replayed in the theater of our minds. The barrier to being able to recall everything we might want to is not that it wasn't "stored" in memory, then; rather, the problem is finding the right cue to access the memory and properly configure our neural circuits. The more we access a memory, the more active become the retrieval and recollection circuits, and the more facile we are with the cues necessary to get at the memory. In theory, if we only had the right cues, we could access any past experience.

Think for a moment of your third-grade teacher—this is probably something you haven't thought about in a long time, but there it is—an instant memory. If you continue to think about your teacher, your class-room, you might be able to recall some other things about third grade such as the desks in the classroom, the hallways of your school, your play-mates. These cues are rather generic and not very vivid. However, if I could show you your third-grade class photo, you might suddenly begin to recall all kinds of things you had forgotten—the names of your class-mates, the subjects you learned in class, the games you played at lunchtime. A song playing comprises a very specific and vivid set of mem-ory cues. Because the multiple-trace memory models assume that con-text is encoded along with memory traces, the music that you have listened to at various times in your life is cross-coded with the events of those times. That is, the music is linked to events of the time, and those events are linked to the music.

A maxim of memory theory is that unique cues are the most effective at bringing up memories; the more items or contexts a particular cue is associated with, the less effective it will be at bringing up a particular memory. This is why, although certain songs may be associated with certain times of your life, they are not very effective cues for retrieving memories from those times if the songs have continued to play all along and you're accustomed to hearing them—as often happens with classic rock stations or the classical radio stations that rely on a somewhat lim-ited repertoire of "popular" classical pieces. But as soon as we hear a song that we haven't heard since a particular time in our lives, the flood-gates of memory open and we're immersed in memories. The song has acted as a unique cue, a key unlocking all the experiences associated with the memory for the song, its time and place. And because memory and categorization are linked, a song can access not just specific memo-ries, but more general, categorical memories. That's why if you hear one 1970s disco song—"YMCA" by the Village People, for example—you might find other songs from that genre playing in your head, such as "I Love the Nightlife" by Alicia Bridges and "The Hustle" by Van McCoy.

Memory affects the music-listening experience so profoundly that it would be not be hyperbole to say that without memory there would be no music. As scores of theorists and philosophers have noted, as well as the songwriter John Hartford in his song "Tryin' to Do Something to Get Your Attention," music is based on repetition. Music works because we remember the tones we have just heard and are relating them to the ones that are just now being played. Those groups of tones—phrases—might come up later in the piece in a variation or transposition that tickles our memory system at the same time as it activates our emotional centers. In the past ten years, neuroscientists have shown just how intimately related our memory system is with our emotional system. The amygdala, long considered the seat of emotions in mammals, sits adjacent to the hippocampus, long considered the crucial structure for memory storage, if not memory retrieval. Now we know that the amygdala is involved in memory; in particular, it is highly activated by any experience or memory that has a strong emotional component. Every neuroimaging study that my laboratory has done has shown amygdala activation to music, but not to random collections of sounds or musical tones. Repetition, when done skillfully by a master composer, is emotionally satisfying to our brains, and makes the listening experience as pleasurable as it is.

6. After Dessert, Crick Was Still Four Seats Away from Me

Music, Emotion, and the Reptilian Brain

A s I've discussed, most music is foot-tapping music. We listen to music that has a pulse, something you can tap your foot to, or at least tap the foot in your mind to. This pulse, with few exceptions, is regular and evenly spaced in time. This regular pulse causes us to expect events to occur at certain points in time. Like the clickety-clack of a railroad track, it lets us know that we're continuing to move forward, that we're in motion, that everything is all right.

Composers sometimes suspend the sense of pulse, such as in the first few measures of Beethoven's Fifth Symphony. We hear "bump-bump-bump-baaaah" and the music stops. We're not sure when we're going to hear a sound again. The composer repeats the phrase—using different pitches—but after that second rest, we're off and running, with a regular foot-tappable meter. Other times, composers give us the pulse explicitly, but then intentionally soften its presentation before coming in with a heavy articulation of it for dramatic effect. "Honky Tonk Women" by the Rolling Stones begins with cowbell, followed by drums, followed by electric guitar; the meter stays the same and our sense of the beat does, too, but the intensity of the strong beats unfolds. (And when we listen on headphones the cowbell comes out of only one ear for more dramatic effect.) This is typical of heavy metal and rock anthems. "Back in Black" by

AC/DC begins with the high-hat cymbal and muted guitar chords that sound almost like a small snare drum for eight beats until the onslaught of electric guitar comes in. Jimi Hendrix does the same thing in opening "Purple Haze"—eight quarter notes on the guitar and bass, single notes that explicitly set up the meter for us before Mitch Mitchell's thunderous drums are ushered in. Sometimes composers tease us, setting up expectations for the meter and then taking them away before settling on something strong—a sort of musical joke that they let us in on. Stevie Wonder's "Golden Lady" and Fleetwood Mac's "Hypnotized" establish a meter that is changed when the rest of the instruments come in. Frank Zappa was a master at this.

Some types of music seem more rhythmically driven than others, of course. Although "Eine Kleine Nachtmusik" and "Stayin' Alive" both have a definable meter, the second one is more likely to make most people get up and dance (at least that's the way we felt in the 1970s). In order to be moved by music (physically and emotionally) it helps a great deal to have a readily predictable beat. Composers accomplish this by subdividing the beat in different ways, and accenting some notes differently than others; a lot of this has to do with performance as well. When we talk about a great groove in music, we're not talking in some jive sixties Austin Powers fab lingo, baby; we're talking about the way in which these beat divisions create a strong momentum. Groove is that quality that moves the song forward, the musical equivalent to a book that you can't put down. When a song has a good groove, it invites us into a sonic world that we don't want to leave. Although we are aware of the pulse of the song, external time seems to stand still, and we don't want the song to ever end.

Groove has to do with a particular performer or particular performance, not with what is written on paper. Groove can be a subtle aspect of performance that comes and goes from one day to another, even with the same group of musicians. And, of course, listeners disagree about whether something has a good groove or not, but to establish some common ground for the topic here, most people feel that "Shout" by the Isley Brothers and "Super Freak" by Rick James have a great groove, as does

"Sledgehammer" by Peter Gabriel. "I'm On Fire" by Bruce Springsteen, "Superstition" by Stevie Wonder, and "Ohio" by the Pretenders all have great grooves, and are very different from one another. But there is no formula for how to create a great one, as every R & B musician who has tried to copy the groove of classic tunes like those by the Temptations and Ray Charles will tell you. The fact that we can point to relatively few songs that have it is evidence that copying it is not so easy.

One element that gives "Superstition" its great groove is Stevie Wonder's drumming. In the opening few seconds of "Superstition," when Stevie's high-hat cymbal is playing alone, you can hear part of the secret to the song's groove. Drummers consider the high-hat to be their time-keeper. Even if listeners can't hear it in a loud passage, the drummer uses it as a point of reference for himself. The beat Stevie plays on the high-hat is never exactly the same way twice; he throws in little extra taps, hits, and rests. Moreover, every note that he plays on the cymbal has a slightly different volume—nuances in his performance that add to the sense of tension. The snare drum starts with bum-(rest)-bum-bum-pa and we're into the high-hat pattern:

DOOT-doot-doot-dootah DOOtah-doot-doot-dootah
DOOT-daat-doot-dootah DOOT-dootah-dootah-doot

The genius of his playing is that he keeps us on our mental toes by changing aspects of the pattern every time he plays it, holding just enough of it the same to keep us grounded and oriented. Here, he plays the same rhythm at the beginning of each line, but changes the rhythm in the second part of the line, in a "call-and-response" pattern. He also uses his skill as a drummer to alter the timbre of his high-hat in one key place: for the second note of the second line, in which he has kept the rhythm the same, he hits the cymbal differently to make it "speak" in a separate voice; if his cymbal were a voice, it's as if he changed the vowel sound that was speaking.

Musicians generally agree that groove works best when it is not strictly metronomic—that is, when it is not perfectly machinelike. Although some danceable songs have been made with drum machines (Michael Jackson's "Billie Jean" and Paula Abdul's "Straight Up," for example), the gold standard of groove is usually a drummer who changes the tempo slightly according to aesthetic and emotional nuances of the music; we say then that the rhythm track, that the drums, "breathe." Steely Dan spent months trying to edit, reedit, shift, push, and pull the drum-machine parts on their album *Two Against Nature* in order to get them to sound as if a human had played them, to balance groove with breathing. But changing local, as opposed to global, tempos like this doesn't change meter, the basic structure of the pulse; it only changes the precise moment that beats will occur, not whether they group in twos, threes, or fours, and not the global pace of the song.

We don't usually talk about groove in the context of classical music, but most operas, symphonies, sonatas, concertos, and string quartets have a definable meter and pulse, which generally corresponds to the conductor's movements; the conductor is showing the musicians where the beats are, sometimes stretching them out or compressing them for emotional communication. Real conversations between people, real pleas of forgiveness, expressions of anger, courtship, storytelling, planning, and parenting don't occur at the precise clips of a machine. To the extent that music is reflecting the dynamics of our emotional lives, and our interpersonal interactions, it needs to swell and contract, to speed up and slow down, to pause and reflect. The only way that we can feel or know these timing variations is if a computational system in the brain has extracted information about when the beats are supposed to occur. The brain needs to create a model of a constant pulse—a schema—so that we know when the musicians are deviating from it. This is similar to variations of a melody: We need to have a mental representation of what the melody is in order to know—and appreciate—when the musician is taking liberties with it.

Metrical extraction, knowing what the pulse is and when we expect it to occur, is a crucial part of musical emotion. Music communicates to us emotionally through systematic violations of expectations. These violations can occur in any domain—the domain of pitch, timbre, contour, rhythm, tempo, and so on—but occur they must. Music is organized sound, but the organization has to involve some element of the unexpected or it is emotionally flat and robotic. Too much organization may technically still be music, but it would be music that no one wants to listen to. Scales, for example, are organized, but most parents get sick of hearing their children play them after five minutes.

What of the neural basis for this metrical extraction? From lesion studies we know that rhythm and metrical extraction aren't neurally related to each other. Patients with damage to the left hemisphere can lose the ability to perceive and produce rhythm, but they can still extract meter, and patients with damage to the right hemisphere have shown the opposite pattern. Both of these are neurally separate from melody processing: Robert Zatorre found that lesions to the right temporal lobe affect the perception of melodies more than lesions to the left; Isabelle Peretz discovered that the right hemisphere of the brain contains a contour processor that in effect draws an outline of a melody and analyzes it for later recognition, and this is dissociable from rhythm and meter circuits in the brain.

As we saw with memory, computer models can help us grasp the inner workings of the brain. Peter Desain and Henkjan Honing of the Netherlands developed a computer model that could extract the beat from a piece of music. It relied mainly on amplitude, the fact that meter is defined by loud versus soft beats occurring at regular intervals of alternation. To demonstrate the effectiveness of their system—and because they recognize the value of showmanship, even in science—they hooked up the output of their system to a small electric motor mounted inside a shoe. Their beat-extraction demonstration actually tapped its foot (or at least a shoe on a metal rod) to real pieces of music. I saw this demonstrated at CCRMA in the mid nineties. It was quite impressive. Spectators (I'm calling us that because the sight of men's size-nine black wingtip shoe hanging from a metal rod and connected via a snake of wires to the computer was quite a spectacle) could give a CD to Desain and Honing, and their shoe would, after a few seconds of "listening," start to tap against a piece of plywood. (When the demonstration was over, Perry Cook went up to them and said, "Very nice work . . . but does it come in brown?")

Interestingly, the Desain and Honing system had some of the same weaknesses that real, live humans do: It would sometimes tap its foot in half time or double time, compared to where professional musicians felt that the beat was. Amateurs do this all the time. When a computerized model makes similar mistakes to a human, it is even better evidence that our program is replicating human thought, or at least the types of computational processes underlying thought.

The cerebellum is the part of the brain that is involved closely with timing and with coordinating movements of the body. The word *cerebellum* derives from the Latin for "little brain," and in fact, it looks like a small brain hanging down underneath your cerebrum (the larger, main part of the brain), right at the back of your neck. The cerebellum has two sides, like the cerebrum, and each is divided into subregions. From phylogenetic studies—studies of the brains of different animals up and down the genetic ladder—we've learned that the cerebellum is one of the oldest parts of the brain, evolutionarily speaking. In popular language, it has sometimes been referred to as the reptilian brain. Although it weighs only 10 percent as much as the rest of the brain, it contains 50 to 80 percent of the total number of neurons. The function of this oldest part of the brain is something that is crucial to music: timing.

The cerebellum has traditionally been thought of as that part of the brain that guides movement. Most movements made by most animals have a repetitive, oscillatory quality. When we walk or run, we tend to do so at a more or less constant pace; our body settles into a gait and we maintain it. When fish swim or birds fly, they tend to flip their fins or flap their wings at a more or less constant rate. The cerebellum is involved in maintaining this rate, or gait. One of the hallmarks of Parkinson's disease is difficulty walking, and we now know that cerebellar degeneration accompanies this disease.

But what about music and the cerebellum? In my laboratory we found strong activations in the cerebellum when we asked people to listen to music, but not when we asked them to listen to noise. The cerebellum appears to be involved in tracking the beat. And the cerebellum has shown up in our studies in another context: when we ask people to listen to music they like versus music they don't like, or familiar music versus unfamiliar music.

Many people, including ourselves, wondered if these cerebellar activations to liking and familiarity were in error. Then, in the summer of 2003, Vinod Menon told me about the work of Harvard professor Jeremy Schmahmann. Schmahmann has been swimming upstream against the tide of traditionalists who said that the cerebellum is for timing and movement and nothing else. But through autopsies, neuroimaging, case studies, and studies of other species, Schmahmann and his followers have amassed persuasive evidence that the cerebellum is also involved in emotion. This would account for why it becomes activated when people listen to music they like. He notes that the cerebellum contains massive connections to emotional centers of the brain—the amygdala, which is involved in remembering emotional events, and the frontal lobe, the part of the brain involved in planning and impulse control. What is the connection between emotion and movement, and why would they both be served by the same brain region, a region found even in snakes and lizards? We don't know for sure, but some informed speculation comes through the very best of sources: the codiscoverers of DNA's structure, James Watson and Francis Crick.

Cold Spring Harbor Laboratory is an advanced, high-tech institution on Long Island, specializing in research on neuroscience, neurobiology, cancer, and—as befits an institution whose director is the Nobel laureate James Watson—genetics. Through SUNY Stony Brook, CSHL offers degrees and advanced training in these fields. A colleague of mine, Amandine Penel, was a postdoctoral fellow there for a couple of years. She had taken her Ph.D. in music cognition in Paris while I was earning mine at the University of Oregon; we knew each other from the annual music cognition conferences. Every so often, CSHL sponsors a workshop, an intensive gathering of scientists who are specialists on a particular topic. These workshops span several days; everyone eats and sleeps at the laboratory, and spends all day together hashing out the chosen scientific problem. The idea behind such a gathering is that if the people who are world experts on the topic—often contentiously holding opposite views—can come to some sort of an agreement about certain aspects of the problem, science can move forward more quickly. The CSHL workshops are famous in genomics, plant genetics, and neurobiology.

I was taken by surprise one day when, buried in between rather mundane e-mails about the undergraduate curriculum committee and final examination schedules at McGill, I saw one inviting me to participate in a four-day workshop at Cold Spring Harbor. Here is what I found in my in-box:

Neural Representation and Processing of Temporal Patterns

How is time represented in the brain? How are complex temporal patterns perceived or produced? Processing of temporal patterns is a fundamental component of sensory and motor function. Given the inherent temporal nature of our interaction with the environment, understanding how the brain processes time is a necessary step towards understanding the brain. We aim to bring together the top psychologists, neuroscientists, and theorists in the world working on these problems. Our goals are twofold: First, we wish to bring together researchers from different fields that share a common focus on timing and would benefit greatly from cross-fertilization of ideas. Second, much significant work to date has been carried out on single-temporal-interval processing. Looking to the future, we wish to learn from these studies while extending the discussion to the processing of temporal patterns that are composed of multiple intervals. Temporal pattern perception is growing as a multi-disciplinary field; we anticipate that this meeting may help to discuss and set a cross-disciplinary research agenda.

At first, I thought that the organizers had made a mistake by including my name on the list. I knew all the names of the invited participants that came with the e-mail. They were the giants in my field—the George Martins and Paul McCartneys, the Seiji Ozawas and Yo-Yo Mas of timing research. Paula Tallal had discovered, with her collaborator Mike Merzenich of UC San Francisco, that dyslexia was related to a timing deficit in children's auditory systems. She had also published some of the most influential fMRI studies of speech and the brain, showing where in the brain phonetic processing occurs. Rich Ivry was my intellectual cousin, one of the brightest cognitive neuroscientists of my generation, who had received his Ph.D. from Steve Keele at the University of Oregon and had done ground-breaking work on the cerebellum and on the cognitive aspects of motor control. Rich has a very low-key, down-to-earth manner, and he can cut to the heart of a scientific issue with razor precision.

Randy Gallistel was a top mathematical psychologist who modeled memory and learning processes in humans and mice; I had read his papers forward and backward. Bruno Repp had been Amandine Penel's first postdoctoral advisor, and had been a reviewer on the first two papers I ever published (the experiments of people singing pop songs very near the correct pitch and tempo). The other world expert on musical timing, Mari Reiss Jones, was also invited. She had done the most important work on the role of attention in music cognition, and had an influential model of how musical accents, meter, rhythm, and expectations converge to create our knowledge of musical structure. And John Hopfield, the inventor of Hopfield nets, one of the most important classes of PDP neural-network models, was going to be there! When I arrived at Cold Spring Harbor, I felt like a girl backstage at a 1957 Elvis concert.

The conference was intense. Researchers there couldn't agree on basic issues, such as how to distinguish an oscillator from a timekeeper, or whether different neural processes were involved in estimating the length of a silent interval, versus the length of a time span that was filled with regular pulses.

As a group, we realized—just as the organizers had hoped—that much of what impeded true progress in the field was that we were using different terminology to mean the same things, and in many cases, we were using a single word (such as *timing*) to mean very different things, and following very different elementary assumptions.

When you hear someone use a word like *planum temporale* (a neural structure), you think he's using it the same way you are. But in science, as in music, assumptions can be the death of you. One person considered that the planum temporale had to be defined anatomically, another that it had to be defined functionally. We argued about the importance of gray matter versus white matter, about what it means to have two events be synchronous—do they actually have to happen at exactly the same time, or just at what appears perceptually to be the same time?

At night, we had catered dinners and lots of beer and red wine, and we continued discussions as we ate and drank. My doctoral student Bradley Vines came down as an observer, and he played saxophone for everyone. I played guitar with a few of the group who were musicians, and Amandine sang.

Because the meeting was about timing, most of the people there hadn't paid much attention to Schmahmann's work or to the possible connection between emotion and the cerebellum. But Ivry had; he knew Schmahmann's work and was intrigued by it. In our discussions, he cast a light on similarities between music perception and motor action planning, which I hadn't been able to see in my own experiment. He agreed that the heart of the mystery of music must involve the cerebellum. When I met Watson, he told me he also felt there to be a plausible connection among the cerebellum, timing, music, and emotion. But what could that connection be? What was its evolutionary basis?

A few months later, I visited my close collaborator Ursula Bellugi at the Salk Institute, in La Jolla, California. The Salk Institute sits on a pristine piece of land overlooking the Pacific Ocean. Bellugi, a student of the great Roger Brown at Harvard in the 1960s, runs the Cognitive Neuroscience Laboratory there. Among many, many "firsts" and landmark findings in her career, she was the first to show that sign language is truly a language (with syntactic structure, it is not just an ad hoc or disorganized bunch of gestures), and she thus showed that Chomsky's linguistic module is not for spoken language only. She also has done ground-breaking work on spatial cognition, gesture, neurodevelopmental disorders, and the ability of neurons to change function—neuroplasticity.

Ursula and I have been working together for ten years to uncover the genetic basis of musicality. What better place was there for the research to be based than an institute headed by Francis Crick, the man who, with Watson, discovered the structure of DNA? I had gone there, as I do every year, so that we could look at our data together, and work on preparing articles for publication. Ursula and I like sitting in the same room together, looking at the same computer screen, where we can point to the chromosome diagrams, look at brain activations, and talk over what they mean for our hypotheses.

Once a week, the Salk Institute had a "professors' lunch" at which venerable scientists sat around a large square table with Francis Crick, the Institute's director. Visitors were seldom allowed; this was a private forum at which scientists felt free to speculate. I'd heard of this hallowed ground and dreamed of visiting it.

In Crick's book *The Astonishing Hypothesis*, he argued that consciousness arises from the brain, that the sum total of our thoughts, beliefs, desires, and feelings comes from the activities of neurons, glial cells, and the molecules and atoms that make them up. This was interesting, but as I've said, I am somewhat biased against mapping the mind for mapping's own sake, and biased toward understanding how the machinery gives rise to human experience.

What really made Crick interesting to me was not his brilliant work on DNA or his stewardship of the Salk Institute, or even *The Astonishing Hypothesis*. It was his book *What Mad Pursuit*, about his early years in science. In fact, it was precisely this passage, because I, too, had begun my scientific career somewhat late in life.

When the war finally came to an end, I was at a loss as to what to do. . . . I took stock of my qualifications. A not-very-good degree, redeemed somewhat by my achievements at the Admiralty. A knowledge of certain restricted parts of magnetism and hydrodynamics, neither of them subjects for which I felt the least bit of enthusiasm. No published papers at all. . . . Only gradually did I re-alize that this lack of qualification could be an advantage. By the time most scientists have reached age thirty they are trapped by their own expertise. They have invested so much effort in one par-ticular field that it is often extremely difficult, at that time in their careers, to make a radical change. I, on the other hand, knew noth-ing, except for a basic training in somewhat old-fashioned physics and mathematics and an ability to turn my hand to new things. . . . Since I essentially knew nothing, I had an almost completely free choice. . . .

Crick's own search had encouraged me to take my lack of experience as a license to think about cognitive neuroscience differently than other people, and it inspired me to reach beyond what seemed to be the shal-low limits of my own grasp.

I drove to Ursula's lab from my hotel one morning to get an early start. "Early" for me was seven A.M., but Ursula had been in the lab since six. While we worked together in her office, typing on our computer key-boards, Ursula put down her coffee and looked at me with a pixielike twinkle in her eye. "Would you like to meet Francis today?" The coinci-dence of my having met Watson, Crick's Nobel laureate twin, only a few months before was striking.

I felt a rush of panic as an old memory assaulted me. When I was just getting started as a record producer, Michelle Zarin, the manager of the top recording studio in San Francisco, the Automatt, would have Friday afternoon wine-and-cheese get-togethers in her office to which only the inner circle were invited. For months as I worked with unknown bands like the Afflicted and the Dimes, I saw rock's royalty file into her office on Friday afternoons: Carlos Santana, Huey Lewis, the producers Jim Gaines and Bob Johnston. One Friday she told me that Ron Nevison was going to be in town—he had engineered my favorite Led Zeppelin records, and had worked with the Who. Michelle led me into her office and showed me where to stand in the semicircle that began to form. People drank and chatted, and I listened respectfully. But Ron Nevison seemed oblivious to me, and he was the one I really wanted to meet. I looked at my watch—fifteen minutes went by. Boz Scaggs (another client) was on the stereo in the corner. "Lowdown." "Lido." Twenty minutes had gone by. Was I ever going to meet Nevison? "We're All Alone" came on, and—as music can sometimes do—the lyrics got under my skin. I had to take matters into my own hands. I walked over to Nevison and introduced myself. He shook my hand and returned to the conversation he was having. That was it. Michelle scolded me later—this sort of thing is simply not done. If I had waited until she introduced me, she would have reminded him that I was the young producer she had spoken to him about, the potential apprentice, the respectful and thoughtful young man that she wanted him to meet. I never saw Nevison again.

At lunchtime, Ursula and I walked out into the warm spring San Diego air. I could hear seagulls calling overhead. We walked to the corner of the Salk campus with the best view of the Pacific, and walked up three flights of stairs to the professors' lunchroom. I immediately recognized Crick, although he looked quite frail—he was in his late eighties, knocking tentatively on ninety's door. Ursula showed me to a seat about four people away from him to his right.

The lunch conversation was a cacophony. I heard snippets of conversations about a cancer gene that one of the professors had just identified, and about decoding the genetics of the visual system in the squid. Someone else was speculating on a pharmaceutical intervention to slow the memory loss associated with Alzheimer's. Crick mostly listened, but he occasionally spoke, in a voice so soft I couldn't hear a word. The lunchroom thinned out as the professors finished eating.

After dessert, Crick was still four seats away from me, animatedly talking to someone on his left, facing away from us. I wanted to meet him, to talk about *The Astonishing Hypothesis*, to find out what he thought about the relationship among cognition, emotion, and motor control. And what did the codiscoverer of DNA's structure have to say about a possible genetic basis for music?

Ursula, sensing my impatience, said that she'd introduce me to Francis on our way out. I was disappointed, anticipating a "hello-goodbye."

Ursula took me by the elbow; she is only four foot ten and has to reach *up* to get to my elbow. She brought me over to Crick, who was talking about leptons and muons with a colleague. She interrupted him. "Francis," she said, "I just wanted to introduce you to my colleague Dan Levitin, from McGill, who works on Williams and music with me." Before Crick could say a word, Ursula pulled me by the elbow toward the door. Crick's eyes lit up. He sat up straight in his chair. "Music," he said. He brushed away his lepton colleague. "I'd like to talk to you about that sometime," he said. "Well," Ursula said slyly, "we have some time right now."

Crick wanted to know if we had done any neuroimaging studies of music; I told him about our studies on music and the cerebellum. He was intrigued by our results, and at the possibility that the cerebellum might be involved in musical emotion. The cerebellum's role in helping performers and conductors keep track of musical time and to maintain a constant tempo was well known. Many also assumed it was involved in keeping track of musical time in listeners. But where did emotion fit in? What might have been the evolutionary connection between emotion, timing, and movement?

To begin with, what might be the evolutionary basis for emotions? Scientists can't even agree about what emotions are. We distinguish between emotions (temporary states that are usually the result of some external event, either present, remembered, or anticipated), moods (not-so-temporary, longer-lasting states that may or may not have an external cause), and traits (a proclivity or tendency to display certain states, such as "She is generally a happy person," or "He never seems satisfied"). Some scientists use the word *affect* to refer to the valence (positive or negative) of our internal states, and reserve the word *emotion* to refer to particular states. Affect can thus take on only two values (or a third value if you count "no affective state") and within each we have a range of emotions: Positive emotions would include happiness and satiety, negative would include fear and anger.

Crick and I talked about how in evolutionary history, emotions were closely associated with motivation. Crick reminded me that emotions for our ancient hominid ancestors were a neurochemical state that served to motivate us to act, generally for survival purposes. We see a lion and that instantly generates fear, an internal state—an emotion—that results when a particular cocktail of neurotransmitters and firing rates is achieved. This state that we call "fear" motivates us to stop what we're doing and—without thinking about it—run. We eat a piece of bad food and we feel the emotion of disgust; immediately certain physiological reflexes kick in, such as a scrunching up of the nose (to avoid letting in a possible toxic odor) and a sticking out of the tongue (to eject the offending food); we also constrict our throat to limit the amount of food that gets into our stomach. We see a body of water after we've been wandering for hours, and we're elated—we drink and the satiety fills us with a sense of well-being and contentment, emotions that cause us to remember where that watering hole is for next time.

Not all emotional activities lead to motor movements, but many of the important ones do, and running is prime among them. We can run faster and far more efficiently if we do so with a regular gait—we're less likely to stumble or lose our balance. The role of the cerebellum is clear here. And the idea that emotions might be bound up with cerebellar neurons make sense too. The most crucial survival activities often involve running—away from a predator or toward escaping prey—and our ancestors needed to react quickly, instantly, without analyzing the situation and studying the best course of action. In short, those of our ancestors who were endowed with an emotional system that was directly connected to their motor system could react more quickly, and thus live to reproduce and pass on those genes to another generation.

What really interested Crick wasn't evolutionary origins of behavior so much as the data. Crick had read the work of Schmahmann, who was attempting to resurrect many old ideas that had fallen into disfavor or had simply been forgotten, such as a 1934 paper suggesting that the cerebellum was involved in the modulation of arousal, attention, and sleep. During the 1970s, we learned that lesions to particular regions of the cerebellum could cause dramatic changes in arousal. Monkeys with a lesion to one portion of their cerebellum would experience rage—called sham rage by scientists because there was nothing in the environment to

cause this reaction. (Of course, the monkeys had every reason to be en-
raged because some surgeon had just lesioned parts of their brains, but
the experiments show that they only exhibit rage after these cerebellar—
but not other—lesions.) Lesions to other parts of the cerebellum cause
calm and have been used clinically to soothe schizophrenics. Electrical
stimulation of a thin strip of tissue at the center of the cerebellum, called
the vermis, can lead to aggression in humans, and in a different region to
a reduction in anxiety and depression.

Crick's dessert plate was still in front of him, and he pushed it away.
He clutched a glass of ice water in his hands. I could see the veins of his
hands through his skin. For a moment I thought I could actually see his
pulse. He became quiet for a moment, staring, thinking. The room was
completely still now, but through an open window we could hear the
crashing of the waves below.

We discussed the work of neurobiologists who had shown in the
1970s that the inner ear doesn't send all of its connections to the auditory
cortex, as was previously believed. In cats and rats, animals whose auditory systems are well known and bear a marked resemblance to our own,
there are projections directly from the inner ear to the cerebellum—
connections that come into the cerebellum from the ear—that coordinate the movements involved in orienting the animal to an auditory
stimulus in space. There are even location-sensitive neurons in the cerebellum, an efficient way of rapidly orienting the head or body to a source.
These areas in turn send projections out to the areas in the frontal lobe
that my studies with Vinod Menon and Ursula found to be active in processing both language and music—regions in the inferior frontal and
orbitofrontal cortex. What was going on here? Why would the connections from the ear bypass the auditory cortex, the central receiving area
for hearing, and send masses of fibers to the cerebellum, a center of motor control (and perhaps, we were learning, of emotion)?

Redundancy and distribution of function are crucial principles of
neuroanatomy. The name of the game is that an organism has to live long
enough to pass on its genes through reproduction. Life is dangerous;
there are a lot of opportunities to get whacked in the head and poten-

tially lose some brain function. To continue to function after a brain injury requires that a blow to a single part of the brain doesn't shut down the whole system. Important brain systems evolved additional, supplementary pathways.

Our perceptual system is exquisitely tuned to detect changes in the environment, because change can be a signal that danger is imminent. We see this in each of the five senses. Our visual system, while endowed with a capacity to see millions of colors and to see in the dark when illumination is as dim as one photon in a million, is most sensitive to sudden change. An entire region of the visual cortex, area MT, specializes in detecting motion; neurons there fire when an object in our visual field moves. We've all had the experience of an insect landing on our neck and we instinctively slap it—our touch system noticed an extremely subtle change in pressure on our skin. And although it is now a staple of children's cartoons, the power of a change in smell—the odor wafting through the air from an apple pie cooling on a neighbor's windowsill—can cause an alerting and orienting reaction in us. But sounds typically trigger the greatest startle reactions. A sudden noise causes us to jump out of our seats, to turn out heads, to duck, or to cover our ears.

The auditory startle is the fastest and arguably the most important of our startle responses. This makes sense: In the world we live in, surrounded by a blanket of atmosphere, the sudden movement of an object—particularly a large one—causes an air disturbance. This movement of air molecules is perceived by us as sound. The principle of redundancy dictates that our nervous system needs to be able to react to sound input even if it becomes partially damaged. The deeper we look inside the brain, the more we find redundant pathways, latent circuits, and connections among systems that we weren't aware of before. These secondary systems serve an important survival function. The scientific literature has recently featured articles on people whose visual pathways were cut, but who can still "see." Although they aren't consciously aware of seeing anything—in fact they claim to be blind—they can still orient toward objects, and in some cases identify them.

A vestigial or supplementary auditory system also appears to be in place involving the cerebellum. This preserves our ability to react quickly—emotionally and with movement—to potentially dangerous sounds.

Related to the startle reflex, and to the auditory system's exquisite sensitivity to change, is the habituation circuit. If your refrigerator has a hum, you get so used to it that you no longer notice it—that is habituation. A rat sleeping in his hole in the ground hears a loud noise above. This could be the footstep of a predator, and he should rightly startle. But it could also be the sound of a branch blowing in the wind, hitting the ground above him more or less rhythmically. If, after one or two dozen taps of the branch against the roof of his house, he finds he is in no danger, he should ignore these sounds, realizing that they are no threat. If the intensity or frequency should change, this indicates that environmental conditions have changed and that he should start to notice. Maybe the wind has picked up and its added velocity will cause the branch to poke through his rodentine residence. Maybe the wind has died down, and it is safe for him to go out and seek food and mates without fear of being blown away by torrential winds. Habituation is an important and necessary process to separate the threatening from the nonthreatening. The cerebellum acts as something of a timekeeper, so when it is damaged, its ability to track the regularity of sensory stimulation is compromised, and habituation goes out the window.

Ursula told Crick of Albert Galaburda's discovery, at Harvard, that individuals with Williams syndrome (WS) have defects in the way their cerebellums form. Williams occurs when about twenty genes turn up missing on one chromosome (chromosome 7). This happens in one out of twenty thousand births, and so it is about one fourth as common as the better-known developmental disorder Down syndrome. Like Down syndrome, Williams results from a mistake in genetic transcription that occurs early in the stages of fetal development. Out of the twenty-five thousand or so genes that we have, the loss of these twenty is devastating. People with Williams can end up with profound intellectual impairment. Few of them learn to count, tell time, or read. Yet, they have more or less intact language skills, they are very musical, and they are unusually outgoing and pleasant; if anything, they are more emotional than the rest of us, and they are certainly more friendly and gregarious than the average person. Making music and meeting new people tend to be two of their favorite things to do. Schmahmann had found that lesions to the cerebellum can create Williams-like symptoms, with people suddenly becoming too outgoing, and acting overly familiar with strangers.

A couple of years ago I was asked to visit a teenage boy with WS. Kenny was outgoing, cheerful, and loved music, but he had an IQ of less than fifty, meaning that at the age of fourteen he had the mental capacity of a seven-year-old. In addition, as with most people struck with Williams syndrome, he had very poor eye-hand coordination, and had difficulty buttoning up his sweater (his mother had to help him), tying his own shoes (he had Velcro straps instead of laces), and he even had difficulty climbing stairs or getting food from his plate to his mouth. But he played the clarinet. There were a few pieces that he had learned, and he was able to execute the numerous and complicated finger movements to play them. He could not name the notes, and couldn't tell me what he was doing at any one point of the piece—it was as though his fingers had a mind of their own. Suddenly the eye-hand coordination problems were gone! But then as soon as he stopped playing, he needed help opening the case to put the clarinet back.

Allan Reiss at Stanford University Medical School has shown that the neocerebellum, the newest part of the cerebellum, is larger than normal in those with WS. Something about movement when it could be entrained to music was different in people with WS than other kinds of movement. Knowing that their cerebellar morphometry was different from others' suggested that the cerebellum might be the part of them that had a "mind of its own," and that could tell us something about how the cerebellum normally influences music processing in people without WS. The cerebellum is central to something about emotion—startle, fear, rage, calm, gregariousness. It was now implicated in auditory processing.

Still sitting with me, long after the lunch plates were cleared, Crick mentioned "the binding problem," one of the most difficult problems in cognitive neuroscience. Most objects have a number of different features that are processed by separate neural subsystems—in the case of visual objects, these might be color, shape, motion, contrast, size, and so on. Somehow the brain has to "bind together" these different, distinct components of perception into a coherent whole. I have described how cognitive scientists believe that perception is a constructive process, but what are the neurons actually doing to bring it all together? We know this is a problem from the study of patients with lesions or particular neuropathic diseases such as Balint's syndrome, in which people can recognize only one or two features of an object but cannot hold them together. Some patients can tell you where an object is in their visual field but not its color, or vice versa. Other patients can hear timbre and rhythm but not melody or vice versa. Isabelle Peretz discovered a patient who has absolute pitch but is tone deaf! He can name notes perfectly, but he cannot sing to save his life.

One solution to the binding problem, Crick proposed, was the synchronous firing of neurons throughout the cortex. Part of the "astonishing hypothesis" of Crick's book was that consciousness emerges from the synchronous firing, at 40 Hz, of neurons in the brain. Neuroscientists had generally considered that the operations of the cerebellum occurred at a "preconscious" level because it coordinates things like running, walking, grasping, and reaching that are normally not under conscious control. There's no reason that the cerebellar neurons can't fire at 40 Hz to contribute to consciousness, he said, although we don't normally attribute humanlike consciousness to those organisms that have only a cerebellum, such as the reptiles. "Look at the connections," Crick said. Crick had taught himself neuroanatomy during his time at Salk, and he had noticed that many researchers in cognitive neuroscience were not adhering to their own founding principles, to use the brain as a constraint for hypotheses; Crick had little patience for such people, and believed that true progress would only be made by people rigorously studying details about brain structure and function.

The lepton colleague was now back, reminding Crick of an impending appointment. We all stood up to leave, and Crick turned to me one last time and repeated, "Look at the connections. . . ." I never saw him again. He died a few months later.

The connection between the cerebellum and music wasn't that hard to see. The Cold Spring Harbor participants were talking about how the frontal lobe—the center of the most advanced cognitions in humans—is connected directly to the cerebellum, the most primitive part of the human brain. The connections run in both directions, with each structure influencing the other. Regions in the frontal cortex that Paula Tallal was studying—those that help us to distinguish precise differences in speech sounds—were also connected to the cerebellum. Ivry's work on motor control showed connections between the frontal lobes, occipital cortex (and the motor strip), and the cerebellum. But there was another player in this neural symphony, a structure deep inside the cortex.

In a landmark study in 1999, Anne Blood, a postdoctoral fellow working with Robert Zatorre at the Montreal Neurological Institute, had shown that intense musical emotion—what her subjects described as "thrills and chills"—was associated with brain regions thought to be involved in reward, motivation, and arousal: the ventral striatum, the amygdala, the midbrain, and regions of the frontal cortex. I was particularly interested in the ventral striatum—a structure that includes the nucleus accumbens—because the nucleus accumbens (NAc) is the center of the brain's reward system, playing an important role in pleasure and addiction. The NAc is active when gamblers win a bet, or drug users take their favorite drug. It is also closely involved with the transmission of opioids in the brain, through its ability to release the neurotransmitter dopamine. Avram Goldstein had shown in 1980 that the pleasure of music listening could be blocked by administering the drug nalaxone, believed to interfere with dopamine in the nucleus accumbens. But the particular type of brain scan that Blood and Zatorre had used, positron emission tomography, doesn't have a high enough spatial resolution to detect whether the small nucleus accumbens was involved. Vinod Menon and I had lots of data collected from the higher-resolution fMRI, and we had the resolving power to pinpoint the nucleus accumbens if it was involved in music listening. But to really nail down the story about how pleasure in the brain occurs in response to music, we'd have to show that the nucleus accumbens was involved at just the right time in a sequence of neural structures that are recruited during music listening. The nucleus accumbens would have to be involved following activation of structures in the frontal lobe that process musical structure and meaning. And in order to know that it was the nucleus accumbens's role as a modulator of dopamine, we would have to figure out a way to show that its activation occurred at the same time as activation of other brain structures that were involved in the production and transmission of dopamine—otherwise, we couldn't argue that the nucleus accumbens involvement was anything more than coincidence. Finally, because so much evidence seemed to point to the cerebellum, which we know to also have dopamine receptors, it would have to show up in this analysis as well.

Menon had just read some papers by Karl Friston and his colleagues about a new mathematical technique, called functional and effective connectivity analysis, that would allow us to address these questions, by revealing the way that different brain regions interact during cognitive operations. These new connectivity analyses would allow us to detect associations between neural regions in music processing that conventional techniques cannot address. By measuring the interaction of one brain region with another—constrained by our knowledge of the anatomical connections between them—the technique would permit us to make a moment-by-moment examination of the neural networks induced by music. This is surely what Crick would have wanted to see. The task was not easy; brain scan experiments produce millions and millions of data points; a single session can take up the entire hard drive on an ordinary computer. Analyzing the data in the standard way—just to see which areas are activated, not the new type of analyses we were proposing—can take months. And there was no "off the shelf" statistical program that would do these new analyses for us. Menon spent two months working through the equations necessary to do these analyses, and when he was done, we reanalyzed the data of people listening to classical music we had collected.

We found exactly what we had hoped. Listening to music caused a cascade of brain regions to become activated in a particular order: first, auditory cortex for initial processing of the components of the sound. Then the frontal regions, such as BA44 and BA47, that we had previously identified as being involved in processing musical structure and expectations. Finally, a network of regions—the mesolimbic system—involved in arousal, pleasure, and the transmission of opioids and the production of dopamine, culminating in activation in the nucleus accumbens. And the cerebellum and basal ganglia were active throughout, presumably supporting the processing of rhythm and meter. The rewarding and reinforcing aspects of listening to music seem, then, to be mediated by increasing dopamine levels in the nucleus accumbens, and by the cerebellum's contribution to regulating emotion through its connections to the frontal lobe and the limbic system. Current neuropsychological theories associate positive mood and affect with increased dopamine levels, one of the reasons that many of the newer antidepressants act on the dopaminergic system. Music is clearly a means for improving people's moods. Now we think we know why.

Music appears to mimic some of the features of language and to convey some of the same emotions that vocal communication does, but in a nonreferential, and nonspecific way. It also invokes some of the same neural regions that language does, but far more than language, music taps into primitive brain structures involved with motivation, reward, and emotion. Whether it is the first few hits of the cowbell on "Honky Tonk Women," or the first few notes of "Sheherazade," computational systems in the brain synchronize neural oscillators with the pulse of the music, and begin to predict when the next strong beat will occur. As the music unfolds, the brain constantly updates its estimates of when new beats will occur, and takes satisfaction in matching a mental beat with a real-in-the-world one, and takes delight when a skillful musician violates that expectation in an interesting way—a sort of musical joke that we're all in on. Music breathes, speeds up, and slows down just as the real world does, and our cerebellum finds pleasure in adjusting itself to stay synchronized.

Effective music—groove—involves subtle violations of timing. Just as the rat has an emotional response to a violation of the rhythm of the branch hitting his house, we have an emotional response to the violation of timing in music that is groove. The rat, with no context for the timing violation, experiences it as fear. We know through culture and experience that music is not threatening, and our cognitive system interprets these violations as a source of pleasure and amusement. This emotional response to groove occurs via the ear–cerebellum–nucleus accumbens–limbic circuit rather than via the ear–auditory cortex circuit. Our response to groove is largely pre- or unconscious because it goes through the cerebellum rather than the frontal lobes. What is remarkable is that all these different pathways integrate into our experience of a single song.

The story of your brain on music is the story of an exquisite orchestration of brain regions, involving both the oldest and newest parts of the human brain, and regions as far apart as the cerebellum in the back of the head and the frontal lobes just behind your eyes. It involves a precision choreography of neurochemical release and uptake between logical prediction systems and emotional reward systems. When we love a piece of music, it reminds us of other music we have heard, and it activates memory traces of emotional times in our lives. Your brain on music is all about, as Francis Crick repeated as we left the lunchroom, connections.

7. What Makes a Musician?

Expertise Dissected

On his album *Songs for Swinging Lovers*, Frank Sinatra is awesomely in control of his emotional expression, rhythm, and pitch. Now, I am not a Sinatra fanatic. I only have a half dozen or so of the more than two hundred albums he's released, and I don't like his movies. Frankly, I find most of his repertoire to be just plain sappy; in everything post-1980, he sounds too cocky. Years ago *Billboard* hired me to review the last album he made, duets with popular singers such as Bono and Gloria Estefan. I panned it, writing that Frank "sings with all the satisfaction of a man who just had somebody killed."

But on *Swinging Lovers*, every note he sings is perfectly placed in time and pitch. I don't mean "perfectly" in the strict, as-notated sense; his rhythms and timing are completely wrong in terms of how the music is written on paper, but they are perfect for expressing emotions that go beyond description. His phrasing contains impossibly detailed and subtle nuances—to be able to pay attention to that much detail, to be able to control it, is something I can't imagine. Try to sing along with any song on *Swinging Lovers*. I've never found anyone who could match his phrasing precisely—it is too nuanced, too quirky, too idiosyncratic.

How do people become expert musicians? And why is that of the millions of people who take music lessons as children, relatively few continue to play music as adults? When they find out what I do for a living, many people tell me that they love music listening, but their music lessons "didn't take." I think they're being too hard on themselves. The chasm between musical experts and everyday musicians that has grown so wide in our culture makes people feel discouraged, and for some reason this is uniquely so with music. Even though most of us can't play basketball like Shaquille O'Neal, or cook like Julia Child, we can still enjoy playing a friendly backyard game of hoops, or cooking a holiday meal for our friends and family. This performance chasm does seem to be cultural, specific to contemporary Western society. And although many people say that music lessons didn't take, cognitive neuroscientists have found otherwise in their laboratories. Even just a small exposure to music lessons as a child creates neural circuits for music processing that are enhanced and more efficient than for those who lack training. Music lessons teach us to listen better, and they accelerate our ability to discern structure and form in music, making it easier for us to tell what music we like and what we don't like.

But what about that class of people that we all acknowledge are true musical experts—the Alfred Brendels, Sarah Changs, Wynton Marsalises, and Tori Amoses? How did they get what most of us don't have, an extraordinary facility to play and perform? Do they have a set of abilities—or neural structures—that are of a totally different sort than the rest of us have (a difference of kind) or do they just have more of the same basic stuff all of us are endowed with (a difference of degree)? And do composers and songwriters have a fundamentally different set of skills than players?

The scientific study of expertise has been a major topic within cognitive science for the past thirty years, and musical expertise has tended to be studied within the context of general expertise. In almost all cases, musical expertise has been defined as technical achievement—mastery of an instrument or of compositional skills. The late Michael Howe, and his collaborators Jane Davidson and John Sloboda, launched an international debate when they asked whether the lay notion of "talent" is scientifically defensible. They assumed the following dichotomy: Either high levels of musical achievement are based on innate brain structures (what we refer to as *talent*) or they are simply the result of training and practice. They define talent as something (1) that originates in genetic structures; (2) that is identifiable at an early stage by trained people who can recognize it even before exceptional levels of performance have been acquired; (3) that can be used to predict who is likely to excel; and (4) that only a minority can be identified as having because if everyone were "talented," the concept would lose meaning. The emphasis on early identification entails that we study the development of skills in children. They add that in a domain such as music, "talent" might be manifested differently in different children.

It is evident that some children acquire skills more rapidly than others: The age of onset for walking, talking, and toilet training vary widely from one child to another, even within the same household. There may be genetic factors at work, but it is difficult to separate out ancillary factors—with a presumably environmental component—such as motivation, personality, and family dynamics. Similar factors can influence musical development and can mask the contributions of genetics to musical ability. Brain studies, so far, haven't been of much use in sorting out the issues because it has been difficult to separate cause from effect. Gottfried Schlaug at Harvard collected brain scans of individuals with absolute pitch (AP) and showed that a region in the auditory cortex—the planum temporale—is larger in the AP people than the non-AP people. This suggests that the planum is involved in AP, but it's not clear if it starts out larger in people who eventually acquire AP, or rather, if the acquisition of AP causes the planum to increase in size. The story is clearer in the areas of the brain that are involved in skilled motor movements. Studies of violin players by Thomas Elbert have shown that the region of the brain responsible for moving the left hand—the hand that requires the most precision in violin playing—increases in size as a result of practice. We do not know yet if the propensity for increase preexists in some people and not others.

The strongest evidence for the talent position is that some people simply acquire musical skills more rapidly than others. The evidence against the talent account—or rather, in favor of the view that practice makes perfect—comes from research on how much training the experts or high achievement people actually do. Like experts in mathematics, chess, or sports, experts in music require lengthy periods of instruction and practice in order to acquire the skills necessary to truly excel. In several studies, the very best conservatory students were found to have practiced the most, sometimes twice as much as those who weren't judged as good.

In another study, students were secretly divided into two groups (not revealed to the students so as not to bias them) based on teachers' evaluations of their ability, or the perception of talent. Several years later, the students who achieved the highest performance ratings were those who had practiced the most, irrespective of which "talent" group they had been assigned to previously. This suggests that practice is the cause of achievement, not merely something correlated with it. It further suggests that talent is a label that we're using in a circular fashion: When we say that someone is talented, we think we mean that they have some innate predisposition to excel, but in the end, we only apply the term retrospectively, after they have made significant achievements.

Anders Ericsson, at Florida State University, and his colleagues approach the topic of musical expertise as a general problem in cognitive psychology involving how humans become experts in general. In other words, he takes as a starting assumption that there are certain issues involved in becoming an expert at *anything*; that we can learn about musical expertise by studying expert writers, chess players, athletes, artists, mathematicians, in addition to musicians.

First, what do we mean by "expert"? Generally we mean that it is someone who has reached a high degree of accomplishment relative to other people. As such, expertise is a social judgment; we are making a statement about a few members of a society relative to a larger population. Also, the accomplishment is normally considered to be in a field that we care about. As Sloboda points out, I may become an expert at folding my arms or pronouncing my own name, but this isn't generally considered the same as becoming, say, an expert at chess, at repairing

Porsches, or being able to steal the British crown jewels without being caught.

The emerging picture from such studies is that ten thousand hours of practice is required to achieve the level of mastery associated with being a world-class expert—in anything. In study after study, of composers, basketball players, fiction writers, ice skaters, concert pianists, chess players, master criminals, and what have you, this number comes up again and again. Ten thousand hours is equivalent to roughly three hours a day, or twenty hours a week, of practice over ten years. Of course, this doesn't address why some people don't seem to get anywhere when they practice, and why some people get more out of their practice sessions than others. But no one has yet found a case in which true world-class expertise was accomplished in less time. It seems that it takes the brain this long to assimilate all that it needs to know to achieve true mastery.

The ten-thousand-hours theory is consistent with what we know about how the brain learns. Learning requires the assimilation and consolidation of information in neural tissue. The more experiences we have with something, the stronger the memory/learning trace for that experience becomes. Although people differ in how long it takes them to consolidate information neurally, it remains true that increased practice leads to a greater number of neural traces, which can combine to create a stronger memory representation. This is true whether you subscribe to multiple-trace theory or any number of variants of theories in the neuro-anatomy of memory: The strength of a memory is related to how many times the original stimulus has been experienced.

Memory strength is also a function of how much we care about the experience. Neurochemical tags associated with memories mark them for importance, and we tend to code as important things that carry with them a lot of emotion, either positive or negative. I tell my students if they want to do well on a test, they have to really care about the material as they study it. Caring may, in part, account for some of the early differences we see in how quickly people acquire new skills. If I really like a particular piece of music, I'm going to want to practice it more, and because I care about it, I'm going to attach neurochemical tags to each aspect of the memory that label it as important: The sounds of the piece, the way I move my fingers, if I'm playing a wind instrument the way that I breathe—all these become part of a memory trace that I've encoded as important.

Similarly, if I'm playing an instrument I like, and whose sound pleases me in and of itself, I'm more likely to pay attention to subtle differences in tone, and the ways in which I can moderate and affect the tonal output of my instrument. It is impossible to overestimate the importance of these factors; caring leads to attention, and together they lead to measurable neurochemical changes. Dopamine, the neurotransmitter associated with emotional regulation, alertness, and mood, is released, and the dopaminergic system aids in the encoding of the memory trace.

Owing to various factors, some people who take music lessons are less motivated to practice; their practice is less effective because of motivational and attentional factors. The ten-thousand-hours argument is convincing because it shows up in study after study across many domains. Scientists like order and simplicity, so if we see a number or a formula that pops up in different contexts, we tend to favor it as an explanation. But like many scientific theories, the ten-thousand-hours theory has holes in it, and it needs to account for counterarguments and rebuttals.

The classic rebuttal to the ten-thousand-hours argument goes something like this: "Well, what about Mozart? I hear that he was composing symphonies at the age of four! And even if he was practicing forty hours a week since the day he was born, that doesn't make ten thousand hours." First, there are factual errors in this account: Mozart didn't begin composing until he was six, and he didn't write his first symphony until he was eight. Still, writing a symphony at age eight is unusual, to say the least. Mozart demonstrated precociousness early in his life. But that is not the same as being an expert. Many children write music, and some even write large-scale works when they're as young as eight. And Mozart had extensive training from his father, who was widely considered to be the greatest living music teacher in all of Europe at the time. We don't know how much Mozart practiced, but if he started at age two and worked thirty-two hours a week at it (quite possible, given his father's reputation as a stern taskmaster) he would have made his ten thousand hours by the age of eight. Even if Mozart hadn't practiced that much, the ten-thousand-hours argument doesn't say that it takes ten thousand hours to write a symphony. Clearly Mozart became an expert eventually, but did the writing of that first symphony qualify him as an expert, or did he attain his level of musical expertise sometime later?

John Hayes of Carnegie Mellon asked just this question. Does Mozart's Symphony no. 1 qualify as the work of a musical expert? Put another way, if Mozart hadn't written anything else, would this symphony strike us as the work of a musical genius? Maybe it really isn't very good, and the only reason we know about it is because the child who wrote it grew up to become Mozart—we have a historical interest in it, but not an aesthetic one. Hayes studied the performance programs of the leading orchestras and the catalog of commercial recordings, assuming that better musical works are more likely to be performed and recorded than lesser works. He found that the early works of Mozart were not performed or recorded very often. Musicologists largely regard them as curiosities, compositions that by no means predicted the expert works that were to follow. Those of Mozart's compositions that are considered truly great are those that he wrote well after he had been at it for ten thousand hours.

As we have seen in the debates about memory and categorization, the truth lies somewhere between the two extremes, a composite of the two hypotheses confronting each other in the nature/nurture debate. To understand how this particular synthesis occurs, and what predictions it makes, we need to look more closely at what the geneticists have to say.

Geneticists seek to find a cluster of genes that are associated with particular observable traits. They assume that if there is a genetic contribution to music, it will show up in families, since brothers and sisters share 50 percent of their genes with one another. But it can be difficult to separate out the influence of genes from the influence of the environment in this approach. The environment includes the environment of the womb: the food that the mother eats, whether she smokes or drinks, and other factors that influence the amount of nutrients and oxygen the fetus receives. Even identical twins can experience very different environments from one another within the womb, based on the amount of space they have, their room for movement, and their position.

Distinguishing genetic from environmental influences on a skill that has a learned component, such as music, is difficult. Music tends to run in families. But a child with parents who are musicians is more likely to receive encouragement for her early musical leanings than a child in a nonmusical household, and siblings of that musically raised child are likely to receive similar levels of support. By analogy, parents who speak French are likely to raise children who speak French, and parents who do not are unlikely to do so. We can say that speaking French "runs in families," but I don't know anyone who would claim that speaking French is genetic.

One way that scientists determine the genetic basis of traits or skills is by studying identical twins, especially those who have been reared apart. The Minnesota twins registry, a database kept by the psychologists David Lykken, Thomas Bouchard, and their colleagues, has followed identical and fraternal twins reared apart and reared together. Because fraternal twins share 50 percent of their genetic material, and identical twins share 100 percent, this allows scientists to tease apart the relative influences of nature versus nurture. If something has a genetic component, we would expect it to show up more often in each individual who is an identical twin than in each who is a fraternal twin. Moreover, we would expect it to show up even when the identical twins have been raised in completely separate environments. Behavioral geneticists look for such patterns and form theories about the heritability of certain traits.

The newest approach looks at gene linkages. If a trait appears to be heritable, we can try to isolate the genes that are linked to that trait. (I don't say "responsible for that trait," because interactions among genes are very complicated, and we cannot say with certainty that a single gene "causes" a trait.) This is complicated by the fact that we can have a gene for something without its being active. Not all of the genes that we have are "turned on," or expressed, at all times. Using gene chip expression profiling, we can determine which genes are and which genes aren't expressed at a given time. What does this mean? Our roughly twenty-five thousand genes control the synthesis of proteins that our bodies and brains use to perform all of our biological functions. They control hair growth, hair color, the creation of digestive fluids and saliva, whether we end up being six feet tall or five feet tall. During our growth spurt around the time of puberty, something needs to tell our body to start growing, and a half dozen years later, something has to tell it to stop. These are the genes, carrying instructions about what to do and how to do it.

Using gene chip expression profiling, I can analyze a sample of your RNA and—if I know what I'm looking for—I can tell whether your growth gene is active—that is, expressed—right now. At this point, the analysis of gene expression in the brain isn't practical because current (and foreseeable) techniques require that we analyze a piece of brain tissue. Most people find that unpleasant.

Scientists studying identical twins who've been reared apart have found remarkable similarities. In some cases, the twins were separated at birth, and not even told of each other's existence. They might have been raised in environments that differed a great deal in geography (Maine versus Texas, Nebraska versus New York), in financial means, and in religious or other cultural values. When tracked down twenty or more years later, a number of astonishing similarities emerged. One woman liked to go to the beach and when she did, she would back into the water; her twin (whom she had never met) did exactly the same thing. One man sold life insurance for a living, sang in his church choir, and wore Lone Star beer belt buckles; so did his completely-separated-from-birth identical twin. Studies like these suggested that musicality, religiosity, and criminality had a strong genetic component. How else could you explain such coincidences?

One alternative explanation is statistical, and can be stated like this: "If you look hard enough, and make enough comparisons, you're going to find some really weird coincidences that don't really mean anything." Take any two random people off the street who have no relationship to one another, except perhaps through their common ancestors Adam and Eve. If you look at enough traits, you're bound to find some in common that aren't obvious. I'm not talking about things like "Oh, my gosh! You breathe the atmosphere too!!" but things like "I wash my hair on Tuesdays and Fridays, and I use an herbal shampoo on Tuesdays—scrubbing with only my left hand, and I don't use a conditioner. On Fridays I use an Australian shampoo that has a conditioner built in. Afterward, I read *The New Yorker* while listening to Puccini." Stories like these suggest that there is an underlying connection between these people, in spite of the scientists' assurances that their genes and environment are maximally dissimilar. But all of us differ from one another in thousands upon thousands of different ways, and we all have our quirks. Once in a while we find co-occurrences, and we're surprised. But from a statistical standpoint, it isn't any more surprising than if I think of a number between one and one hundred and you guess it. You may not guess it the first time, but if we play the game long enough, you're going to guess it once in a while (1 percent of the time, to be exact).

A second alternative explanation is social psychological—the way someone looks influences the way that others treat him (with "looks" assumed to be genetic); in general, an organism is acted on by the world in particular ways as a function of its appearance. This intuitive notion has a rich tradition in literature, from Cyrano de Bergerac to Shrek: Shunned by people who were repulsed by their outward appearance, they rarely had the opportunity to show their inner selves and true nature. As a culture we romanticize stories like these, and feel a sense of tragedy about a good person suffering for something he had nothing to do with: his looks. It works in the opposite way as well: good-looking people tend to make more money, get better jobs, and report that they are happier. Even apart from whether someone is considered attractive or not, his appearance affects how we relate to him. Someone who was born with facial features that we associate with trustworthiness—large eyes, for example, with raised eyebrows—is someone people will tend to trust. Someone tall may be given more respect than someone short. The series of encounters we have over our lifetimes are shaped to some extent by the way others see us.

It is no wonder, then, that identical twins may end up developing similar personalities, traits, habits, or quirks. Someone with downturned eyebrows might always look angry, and the world will treat them that way. Someone who looks defenseless will be taken advantage of; someone who looks like a bully may spend a lifetime being asked to fight, and eventually will develop an aggressive personality. We see this principle at work in certain actors. Hugh Grant, Judge Reinhold, Tom Hanks, and Adrien Brody have innocent-looking faces; without doing anything, Grant has an "awww, shucks" look, a face that suggests he has no guile or deceit. This line of reasoning says that some people are born with particular features, and their personalities develop in large part as a reflection of how they look. Genes here are influencing personality, but only in an indirect, secondary way.

It is not difficult to imagine a similar argument applying to musicians, and in particular to vocalists. Doc Watson's voice sounds completely sincere and innocent; I don't know if he is that way in person, and at one level it doesn't matter. It's possible that he became the successful artist he is because of how people react to the voice that he was born with. I'm not talking about being born with (or acquiring) a "great" voice, like Ella Fitzgerald's or Placido Domingo's, I'm talking about expressiveness apart from whether the voice itself is a great instrument. Sometimes as Aimee Mann sings, I hear the traces of a little girl's voice, a vulnerable innocence that moves me because I feel that she is reaching down deep inside and confessing feelings that normally are expressed only to a close friend. Whether she intends to convey this, or really feels this, I don't know—she may have been born with a vocal quality that makes listeners invest her with those feelings, whether she is experiencing them or not. In the end, the essence of music performance is being able to convey emotion. Whether the artist is feeling it or was born with an ability to sound as if she's feeling it may not be important.

I don't mean to imply that the actors and musicians I've mentioned don't have to work at what they do. I don't know any successful musicians who haven't worked hard to get where they are; I don't know any who had success fall into their laps. I've known a lot of artists whom the press has called "overnight sensations," but who spent five or ten years becoming that! Genetics are a starting point that may influence personality or career, or the specific choices one makes in a career. Tom Hanks is a great actor, but he's not likely to get the same kinds of roles as Arnold Schwarzenegger, largely owing the differences in their genetic endowments. Schwarzenegger wasn't born with a body-builder's body; he worked very hard at it, but he had a genetic predisposition toward it. Similarly, being six ten creates a predisposition toward becoming a basketball player rather than a jockey. But it is not enough for someone who is six ten to simply stand on the court—he needs to learn the game and practice for years to become an expert. Body type, which is largely (though not exclusively) genetic, creates predispositions for basketball as it does for acting, dancing, and music.

Musicians, like athletes, actors, dancers, sculptors and painters, use their bodies as well as their minds. The role of the body in the playing of a musical instrument or in singing (less so, of course, in composing and arranging) means that genetic predispositions can contribute strongly to the choice of instruments a musician can play well—and to whether a person chooses to become a musician.

When I was six years old, I saw the Beatles on *The Ed Sullivan Show*, and in what has become a cliché for people of my generation, I decided then that I wanted to play the guitar. My parents, who were of the old school, did not view the guitar as a "serious instrument" and told me to play the family piano instead. But I wanted desperately to play. I would cut out pictures of classical guitarists like Andrés Segovia from magazines and casually leave them around the house. At six, I was still speaking with a prominent lisp that I had had all my life; I didn't get rid of it until age ten when I was embarrassingly plucked out of my fourth-grade class by the public-school speech therapist who spent a grueling two years (at three hours a week) teaching me to change the way that I said the letter *s*. I pointed out that the Beatles must be therious to share the stage of *The Ed Sullivan Show* with such therious artithts as Beverly Thills, Rodgers and Hammerthtein, and John Gielgud. I was relentless.

By 1965, when I was eight, the guitar was everywhere. With San Francisco just fifteen miles away, I could feel a cultural and musical revolution going on, and the guitar was at the center of it all. My parents were still not enthusiastic about me studying the guitar, perhaps because of its association with hippies and drugs, or perhaps as a result of my failure the previous year to practice the piano diligently. I pointed out that by now, the Beatles had been on *The Ed Sullivan Show* four times and my parents finally quasi-relented, agreeing to ask a friend of theirs for advice. "Jack King plays the guitar," my mother said at dinner one night to my father. "We could ask him if he thinks Danny is old enough to begin guitar lessons." Jack, an old college friend of my parents, dropped by the house one day on his way home from work. His guitar sounded different from the ones that had mesmerized me on television and radio; it was a classical guitar, not made for the dark chords of rock and roll. Jack was a big man with large hands, and a short black crew cut. He held the guitar in his arms as one might cradle a baby. I could see the intricate patterns of wood grain bending around the curves of the instrument. He played something for us. He didn't let me touch the guitar, instead he asked me to hold my hand out, and he pressed his palm against mine. He didn't talk to me or look at me, but what he said to my mother I can still hear clearly: "His hands are too small for the guitar."

I now know about three-quarter size and half-size guitars (I even own one), and about Django Reinhardt, one of the greatest guitarists of all time, who had only two fingers on his left hand. But to an eight-year-old, the words of adults can seem unbreakable. By 1966, when I had grown some, and the Beatles were egging me on with electric guitar strains of "Help," I was playing the clarinet and happy to at least be making music. I finally bought my first guitar when I was sixteen and with practice, I learned to play reasonably well; the rock and jazz that I play don't require the long reach that classical guitar does. The very first song I learned to play—in what has become another cliché for my generation—was Led Zeppelin's "Stairway to Heaven" (hey, it was the seventies).

Some musical parts that guitarists with different hands can play will always be difficult for me, but that is always the case with every instrument. On Hollywood Boulevard in Hollywood, California, some of the great rock musicians have placed their handprints in the cement. I was surprised last summer when I put my hands in the imprint left by Jimmy Page (of Led Zeppelin), one of my favorite guitarists, that his hands were no bigger than mine.

Some years ago I shook hands with Oscar Peterson, the great jazz pianist. His hands were very large; the largest hands I have ever shaken, at least twice the size of my own. He began his career playing stride piano, a style dating back to the 1920s in which the pianist plays an octave bass with his left hand and the melody with his right. To be a good stride player, you need to be able to be able to reach keys that are far apart with a minimum of hand movements, and Oscar can stretch a whopping octave and a half with one hand! Oscar's style is related to the kinds of chords he is able to play, chords that someone with smaller hands could not. If Oscar Peterson had been forced to play violin as a child it would have been impossible with those large hands; his wide fingers would make it difficult to play a semitone on the relatively small neck of the violin.

Some people have a biological predisposition toward particular instruments, or toward singing. There may also be a cluster of genes that work together to create the component skills that one must have to become a successful musician: good eye-hand coordination, muscle control, motor control, tenacity, patience, memory for certain kinds of structures and patterns, a sense of rhythm and timing. To be a good musician, one must have these things. Some of these skills are involved in becoming a great anything, especially determination, self-confidence, and patience.

We also know that, on average, successful people have had many more failures than unsuccessful people. This seems counterintuitive. How could successful people have failed more often than everyone else? Failure is unavoidable and sometimes happens randomly. It's what you do after the failure that is important. Successful people have a stick-to-it-iveness. They don't quit. From the president of FedEx to the novelist

Jerzy Kosinsky, from van Gogh to Bill Clinton to Fleetwood Mac, successful people have had many, many failures, but they learn from them and keep going. This quality might be partly innate, but environmental factors must also play a role.

The best guess that scientists currently have about the role of genes and the environment in complex cognitive behaviors is that each is responsible for about 50 percent of the story. Genes may transmit a propensity to be patient, to have good eye-hand coordination, or to be passionate, but certain life events—life events in the broadest sense, meaning not just your conscious experiences and memories, but the food you ate and the food your mother ate while you were in her womb—can influence whether a genetic propensity will be realized or not. Early life traumas, such as the loss of a parent, or physical or emotional abuse, are only the obvious examples of environmental influences causing a genetic predisposition to become either heightened or suppressed. Because of this interaction, we can only make predictions about human behavior at the level of a population, not an individual. In other words, if you know that someone has a genetic predisposition toward criminal behavior, you can't make any predictions about whether he will end up in jail in the next five years. On the other hand, knowing that a hundred people have this predisposition, we can predict that some percentage of them will probably wind up in jail; we simply don't know which ones. And some will never get into any trouble at all.

The same applies to musical genes we may find someday. All we can say is that a group of people with those genes is more likely to produce expert musicians, but we cannot know which individuals will become the experts. This, however, assumes that we'll be able to identify the genetic correlates of musical expertise, and that we can agree on what constitutes musical expertise. Musical expertise has to be about more than strict technique. Music listening and enjoyment, musical memory, and how engaged with music a person is are also aspects of a musical mind and a musical personality. We should take as inclusive an approach as possible in identifying musicality, so as not to exclude those who, while musical in the broad sense, are perhaps not so in a narrow, technical sense. Many of our greatest musical minds weren't considered experts in a technical sense. Irving Berlin, one of the most successful composers of the twentieth century, was a lousy instrumentalist and could barely play the piano.

Even among the elite, top-tier classical musicians, there is more to being a musician than having excellent technique. Both Arthur Rubinstein and Vladimir Horowitz are widely regarded as two of the greatest pianists of the twentieth century but they made mistakes—little technical mistakes—surprisingly often. A wrong note, a rushed note, a note that isn't fingered properly. But as one critic wrote, "Rubinstein makes mistakes on some of his records, but I'll take those interpretations that are filled with passion over the twenty-two-year-old technical wizard who can play the notes but can't convey the meaning."

What most of us turn to music for is an emotional experience. We aren't studying the performance for wrong notes, and so long as they don't jar us out of our reverie, most of us don't notice them. So much of the research on musical expertise has looked for accomplishment in the wrong place, in the facility of fingers rather than the expressiveness of emotion. I recently asked the dean of one of the top music schools in North America about this paradox: At what point in the curriculum is emotion and expressivity taught? Her answer was that they aren't taught. "There is so much to cover in the approved curriculum," she explained, "repertoire, ensemble, and solo training, sight singing, sight reading, music theory—that there simply isn't time to teach expressivity." So how do we get expressive musicians? "Some of them come in already knowing how to move a listener. Usually they've figured it out themselves somewhere along the line." The surprise and disappointment in my face must have been obvious. "Occasionally," she added, almost in a whisper, "if there's an exceptional student, there's time during the last part of their last semester here to coach them on emotion. . . . Usually this is for people who are already performing as soloists in our orchestra, and we help them to coax out more expressivity from their performance." So, at one of the best music schools we have, the raison d'être for music is taught to a select few, and then, only in the last few weeks of a four- or five-year curriculum.

Even the most uptight and analytic among us expect to be moved by Shakespeare and Bach. We can marvel at the craft these geniuses have mastered, a facility with language or with notes, but ultimately that facility must be brought into service for a different type of communication. Jazz fans, for example, are especially demanding of their post-big-band-era heroes, starting with the Miles Davis/John Coltrane/Bill Evans era. We say of lesser jazz musicians who appear detached from their true selves and from emotion that their playing is nothing more than "shucking and jiving," attempts to please the audience through musical obsequies rather than through soul.

So—in a scientific sense—why are some musicians superior to others when it comes to the emotional (versus the technical) dimension of music? This is the great mystery, and no one knows for sure. Musicians haven't yet performed with feeling inside brain scanners, due to technical difficulties. (The scanners we currently use require the subject to stay perfectly still, so as not to blur the brain image; this may change in the coming five years.) Interviews with, and diary entries of, musicians ranging from Beethoven and Tchaikovsky to Rubinstein and Bernstein, B. B. King, and Stevie Wonder suggest that part of communicating emotion involves technical, mechanical factors, and part of it involves something that remains mysterious.

The pianist Alfred Brendel says he doesn't think about notes when he's onstage; he thinks about creating an experience. Stevie Wonder told me in 1996 that when he's performing, he tries to get himself into the same frame of mind and "frame of heart" that he was in when he wrote the song; he tries to capture the same feelings and sentiment, and that helps him to deliver the performance. What this means in terms of how he sings or plays differently is something no one knows. From a neuroscientific perspective, though, this makes perfect sense. As we've seen, remembering music involves setting the neurons that were originally active in the perception of a piece of music back to their original state—reactivating their particular pattern of connectivity, and getting the firing rates as close as possible to their original levels. This means recruiting neurons in the hippocampus, amygdala, and temporal lobes in a neural symphony orchestrated by attention and planning centers in the front lobe.

The neuroanatomist Andrew Arthur Abbie speculated in 1934 a linkage between movement, the brain, and music that is only now becoming proven. He wrote that pathways from the brain stem and cerebellum to the frontal lobes are capable of weaving all sensory experience and accurately coordinated muscular movements into a "homogeneous fabric" and that when this occurs, the result is "man's highest powers as expressed . . . in art." His idea of this neural pathway was that it is dedicated to motor movements that incorporate or reflect a creative purpose. New studies by Marcelo Wanderley of McGill, and by my former doctoral student Bradley Vines (now at Harvard) have shown that nonmusician listeners are exquisitely sensitive to the physical gestures that musicians make. By watching a musical performance with the sound turned off, and attending to things like the musician's arm, shoulder, and torso movements, ordinary listeners can detect a great deal of the expressive intentions of the musician. Add in the sound, and an emergent quality appears—an understanding of the musician's expressive intentions that goes beyond what was available in the sound or the visual image alone.

If music serves to convey feelings through the interaction of physical gestures and sound, the musician needs his brain state to match the emotional state he is trying to express. Although the studies haven't been performed yet, I'm willing to bet that when B.B. is playing the blues and when he is feeling the blues, the neural signatures are very similar. (Of course there will be differences, too, and part of the scientific hurdle will be subtracting out the processes involved in issuing motor commands and listening to music, versus just sitting on a chair, head in hands, and feeling down.) And as listeners, there is every reason to believe that some of our brain states will match those of the musicians we are listening to. In what is a recurring theme of your brain on music, even those of us who lack explicit training in music theory and performance have musical brains, and are expert listeners.

In understanding the neurobehavioral basis of musical expertise and why some people become better performers than others, we need to consider that musical expertise takes many forms, sometimes technical (involving dexterity) and sometimes emotional. The ability to draw us into a performance so that we forget about everything else is also a special kind of ability. Many performers have a personal magnetism, or charisma, that is independent of any other abilities they may or may not have. When Sting is singing, we can't take our ears off of him. When Miles Davis is playing the trumpet, or Eric Clapton the guitar, an invisible force seems to draw us toward him. This doesn't have to do so much with the actual notes they're singing or playing—any number of good musicians can play or sing those notes, perhaps even with better technical facility. Rather, it is what record company executives call "star quality." When we say of a model that she is photogenic, we're talking about how this star quality manifests itself in photographs. The same thing is true for musicians, and how their quality comes across on records— I call this phonogenic.

It is also important to distinguish celebrity from expertise. The factors that contribute to celebrity could be different from, maybe wholly unrelated to, those that contribute to expertise. Neil Young told me that he did not consider himself to be especially talented as a musician, rather, he was one of the lucky ones who managed to become commercially successful. Few people get to pass through the turnstiles of a deal with a major record label, and fewer still maintain careers for decades as Neil has done. But Neil, along with Stevie Wonder and Eric Clapton, attributes a lot of his success not to musical ability but to a good break. Paul Simon agrees. "I've been lucky to have been able to work with some of the most amazing musicians in the world," he said, "and most of them are people no one's ever heard of."

Francis Crick turned his lack of training into a positive aspect of his life's work. Unbound by scientific dogma, he was free—completely free, he wrote—to open his mind and discover science. When an artist brings this freedom, this tabula rasa, to music, the results can be astounding.

Many of the greatest musicians of our era lacked formal training, including Sinatra, Louis Armstrong, John Coltrane, Eric Clapton, Eddie Van Halen, Stevie Wonder, and Joni Mitchell. And in classical music, George Gershwin, Mussorgsky, and David Helfgott are among those who lacked formal training, and Beethoven considered his own training to have been poor according to his diaries.

Joni Mitchell had sung in choirs in public school, but had never taken guitar lessons or any other kind of music lessons. Her music has a unique quality that has been variously described as avant-garde, ethereal, and as bridging classical, folk, jazz, and rock. Joni uses a lot of alternate tunings; that is, instead of tuning the guitar in the customary way, she tunes the strings to pitches of her own choosing. This doesn't mean that she plays notes that other people don't—there are still only twelve notes in a chromatic scale—but it does mean that she can easily reach with her fingers combinations of notes that other guitarists can't reach (regardless of the size of their hands).

An even more important difference involves the way the guitar makes sound. Each of the six strings of the guitar is tuned to a particular pitch. When a guitarist wants a different one, of course, she presses one or more strings down against the neck; this makes the string shorter, which causes it to vibrate more rapidly, making a tone with a higher pitch. A string that is pressed on ("fretted") has a different sound from one that isn't, due to a slight deadening of the string caused by the finger; the unfretted or "open" strings have a clearer, more ringing quality, and they will keep on sounding for a longer time than the ones that are fretted. When two or more of these open strings are allowed to ring together, a unique timbre emerges. By retuning, Joni changed the configuration of which notes are played when a string is open, so that we hear notes ringing that don't usually ring on the guitar, and in combinations we don't usually hear. You can hear it on her songs "Chelsea Morning" and "Refuge of the Roads" for example.

But there is something more to it than that—lots of guitarists use their own tunings, such as David Crosby, Ry Cooder, Leo Kottke, and

Jimmy Page. One night, when I was having dinner with Joni in Los Angeles, she started talking about bass players that she had worked with. She has worked with some of the very best of our generation: Jaco Pastorius, Max Bennett, Larry Klein, and she wrote an entire album with Charles Mingus. Joni will talk compellingly and passionately about alternate tunings for hours, comparing them to the different colors that van Gogh used in his paintings.

While we were waiting for the main course, she went off on a story about how Jaco Pastorius was always arguing with her, challenging her, and generally creating mayhem backstage before they would go on. For example when the first Roland Jazz Chorus amplifier was hand-delivered by the Roland Company to Joni to use at a performance, Jaco picked it up, and moved it over to his corner of the stage. "It's mine," he growled. When Joni approached him, he gave her a fierce look. And that was that.

We were well into twenty minutes of bass-player stories. Because I was a huge fan of Jaco when he played with Weather Report, I interrupted and asked what it was like musically to play with him. She said that he was different from any other bass player she had every played with; that he was the only bass player up to that time that she felt really understood what she was trying to do. That's why she put up with his aggressive behaviors.

"When I first started out," she said, "the record company wanted to give me a producer, someone who had experience churning out hit records. But [David] Crosby said, 'Don't let them—a producer will ruin you. Let's tell them that I'll produce it for you; they'll trust me.' So basically, Crosby put his name as producer to keep the record company out of my way so that I could make the music the way that I wanted to.

"But then the musicians came in and they all had ideas about how they wanted to play. On *my* record! The worst were the bass players because they always wanted to know what the root of the chord was." The "root" of a chord, in music theory, is the note for which the chord is named and around which it is based. A "C major" chord has the note C as its root, for example, and an "E-flat minor" chord has the note E-flat as its root. It is that simple. But the chords Joni plays, as a consequence of her unique composition and guitar-playing styles, aren't typical chords: Joni throws notes together in such a way that the chords can't be easily labeled. "The bass players wanted to know the root because that's what they've been taught to play. But I said, 'Just play something that sounds good, don't worry about what the root is.' And they said, 'We can't do that—we have to play the root or it won't sound right.'"

Because Joni hadn't had music theory and didn't know how to read music, she couldn't tell them the root. She had to tell them what notes she was playing on the guitar, one by one, and they had to figure it out for themselves, painstakingly, one chord at a time. But here is where psychoacoustics and music theory collide in an explosive conflagration: The standard chords that most composers use—C major, E-flat minor, D7, and so on—are unambiguous. No competent musician would need to ask what the root of a chord like those is; it is obvious, and there is only one possibility. Joni's genius is that she creates chords that are ambiguous, chords that could have two or more different roots. When there is no bass playing along with her guitar (as in "Chelsea Morning" or "Sweet Bird"), the listener is left in a state of expansive aesthetic possibilities. Because each chord could be interpreted in two or more different ways, any prediction or expectation that a listener has about what comes next is less grounded in certainty than with traditional chords. And when Joni strings together several of these ambiguous chords, the harmonic complexity greatly increases; each chord sequence can be interpreted in dozens of different ways, depending on how each of its constituents is heard. Since we hold in immediate memory what we've just heard and integrate it with the stream of new music arriving at our ears and brains, attentive listeners to Joni's music—even nonmusicians—can write and rewrite in their minds a multitude of musical interpretations as the piece unfolds; and each new listening brings a new set of contexts, expectations, and interpretations. In this sense, Joni's music is as close to impressionist visual art as anything I've heard.

As soon as a bass player plays a note, he fixes one particular musical interpretation, thus ruining the delicate ambiguity the composer has so artfully constructed. All of the bass players Joni worked with before Jaco insisted on playing roots, or what they perceived to be roots. The brilliance of Jaco, Joni said, is that he instinctively knew to wander around the possibility space, reinforcing the different chord interpretations with equal emphasis, sublimely holding the ambiguity in a delicate, suspended balance. Jaco allowed Joni to have bass guitar on her songs without destroying one of their most expansive qualities. This, then, we figured out at dinner that night, was one of the secrets of why Joni's music sounds unlike anyone else's—its harmonic complexity born out of her strict insistence that the music not be anchored to a single harmonic interpretation. Add in her compelling, phonogenic voice, and we become immersed in an auditory world, a soundscape unlike any other.

Musical memory is another aspect of musical expertise. Many of us know someone who remembers all kinds of details that the rest of us can't. This could be a friend who remembers every joke he's ever heard in his life, while some of us can't even retell one we've heard that same day. My colleague Richard Parncutt, a well-known musicologist and music cognition professor at the University of Graz in Vienna, used to play piano in a tavern to earn money for graduate school. Whenever he comes to Montreal to visit me he sits down at the piano in my living room and accompanies me while I sing. We can play together for a long time: Any song I name, he can play from memory. He also knows the different versions of songs: If I ask him to play "Anything Goes," he'll ask if I want the version by Sinatra, Ella Fitzgerald, or Count Basie! Now, I can probably play or sing a hundred songs from memory. That is typical for someone who has played in bands or orchestras, and who has performed. But Richard seems to know thousands and thousands of songs, both the chords and lyrics. How does he do it? Is it possible for mere memory mortals like me to learn to do this too?

When I was in music school, at the Berklee College of Music in Boston, I ran into someone with an equally remarkable form of musical memory, but different from Richard's. Carla could recognize a piece of music within just three or four seconds and name it. I don't actually know how good she was at singing songs from memory, because we were always busy trying to come up with a melody to stump her, and this was hard to do. Carla eventually took a job at the American Society of Composers and Publishers (ASCAP), a composers' rights organization that monitors radio station playlists in order to collect royalties for ASCAP members. ASCAP workers sit in a room in Manhattan all day, listening to excerpts from radio programs all over the country. To be efficient at their job, and indeed to be hired in the first place, they have to be able to name a song and the performer within just three to five seconds before writing it down in the log and moving on to the next one.

Earlier, I mentioned Kenny, the boy with Williams syndrome who plays the clarinet. Once when Kenny was playing "The Entertainer" (the theme song from *The Sting*), by Scott Joplin, he had difficulty with a certain passage. "Can I try that again?" he asked me, with an eagerness to please that is typical of Williams syndrome. "Of course," I said. Instead of backing up just a few notes or a few seconds in the piece, however, he went all the way back to the beginning! I had seen this before, in recording studios, with master musicians from Carlos Santana to the Clash—a tendency to go back, if not to the beginning of the entire piece, to the beginning of a phrase. It is as though the musician is executing a memorized sequence of muscle movements, and the sequence has to begin from the beginning.

What do these three demonstrations of memory for music have in common? What is going on in the brains of someone with a fantastic musical memory like Richard and Carla, or the "finger memory" that Kenny has? How might those operations be different from—or similar to—the normal neural processes in someone with a merely ordinary musical memory? Expertise in any domain is characterized by a superior memory, but only for things within the domain of expertise. My friend Richard doesn't have a superior memory for everything in life—he still loses his keys just like anyone else. Grandmaster chess players have memorized thousands of board and game configurations. However, their exceptional memory for chess extends only to legal positions of the chess pieces. Asked to memorize random arrangements of pieces on a board, they do no better than novices; in other words, their knowledge of chess-piece positions is schematized, and relies on knowledge of the legal moves and positions that pieces can take. Likewise, experts in music rely on their knowledge of musical structure. Expert musicians excel at remembering chord sequences that are "legal" or make sense within the harmonic systems that they have experience with, but they do no better than anyone else at learning sequences of random chords.

When musicians memorize songs, then, they are relying on a structure for their memory, and the details fit into that structure. This is an efficient and parsimonious way for the brain to function. Rather than memorizing every chord or every note, we build up a framework within which many different songs can fit, a mental template that can accommodate a large number of musical pieces. When learning to play Beethoven's "Pathé-tique" Sonata, the pianist can learn the first eight measures and then, for the next eight, simply needs to know that the same theme is repeated but an octave higher. Any rock musician can play "One After 909" by the Beatles even if he's never played it before, if he is simply told that it is a "standard sixteen-bar blues progression." That phrase is a framework within which thousands of songs fit. "One After 909" has certain nuances that constitute variations of the framework. The point is that musicians don't typically learn new pieces one note at a time once they have reached a certain level of experience, knowledge, and proficiency. They can scaffold on the previous pieces they know, and just note any variations from the standard schema.

Memory for playing a musical piece therefore involves a process very much like that for music listening as we saw in Chapter 4, through establishing standard schemas and expectation. In addition, musicians use chunking, a way of organizing information similar to the way chess players, athletes, and other experts organize information. *Chunking* refers to the process of tying together units of information into groups, and remembering the group as a whole rather than the individual pieces. We do this all the time without much conscious awareness when we have to remember someone's long-distance phone number. If you're trying to remember the phone number of someone in New York City—and if you know other NYC phone numbers and are familiar with them—you don't have to remember the area code as three individual numerals, rather, you remember it as a single unit: 212. Likewise, you may know that Los Angeles is 213, Atlanta is 404, or that the country code for England is 44. The reason that chunking is important is because our brains have limits on how much information they can actively keep track of. There is no practical limit to long-term memory that we know of, but working memory—the contents of our present awareness—is severely limited, generally to nine pieces of information. Encoding a North American phone number as the area code (one unit of information) plus seven digits helps us to avoid that limit. Chess players also employ chunking, remembering board configurations in terms of groups of pieces arranged in standard, easy-to-name patterns.

Musicians also use chunking in several ways. First, they tend to encode in memory an entire chord, rather than the individual notes of the chord; they remember "C major 7" rather than the individual tones C - E - G - B, and they remember the rule for constructing chords, so that they can create those four tones on the spot from just one memory entry. Second, musicians tend to encode sequences of chords, rather than isolated chords. "Plagal cadence," "aeolian cadence," "twelve-bar minor blues with a V-I turnaround," or "rhythm changes" are shorthand labels that musicians use to describe sequences of varying lengths. Having stored the information about what these labels mean allows the musician to recall big chunks of information from a single memory entry. Third, we obtain knowledge as listeners about stylistic norms, and as players about how to produce these norms. Musicians know how to take a song and apply this knowledge—schemas again—to make the song sound like salsa, or grunge, or disco, or heavy metal; each genre and era has stylistic tics or characteristic rhythmic, timbral, or harmonic elements that define it. We can encode those in memory holistically, and then retrieve these features all at once.

These three forms of chunking are what Richard Parncutt uses when he sits at the piano to play thousands of songs. He also knows enough music theory and is acquainted enough with different styles and genres that he can fake his way through a passage he doesn't really know, just as an actor might substitute words that aren't in the script if she momentarily forgets her lines. If Richard is unsure of a note or chord, he'll replace it with one that is stylistically plausible.

Identification memory—the ability that most of us have to identify pieces of music that we've heard before—is similar to memory for faces, photos, even tastes and smells, and there is individual variability, with some people simply being better than others; it is also domain specific, with some people—like my classmate Carla—being especially good at music, while others excel in other sensory domains. Being able to rapidly retrieve a familiar piece of music from memory is one skill, but being able to then quickly and effortlessly attach a label to it, such as the song title, artist, and year of recording (which Carla could do) involves a separate cortical network, which we now believe involves the planum temporale (a structure associated with absolute pitch) and regions of the inferior prefrontal cortex that are known to be required for attaching verbal labels to sensory impressions. Why some people are better at this than others is still unknown, but it may result from an innate or hardwired predisposition in the way their brains formed, and this in turn may have a partial genetic basis.

When learning sequences of notes in a new musical piece, musicians sometimes have to resort to the brute-force approach that most of us took as children in learning new sequences of sounds, such as the alphabet, the U.S. Pledge of Allegiance, or the Lord's Prayer: We simply do everything we can to memorize the information by repeating it over and over again. But this rote memorization is greatly facilitated by a hierarchical organization of the material. Certain words in a text or notes in a musical piece (as we saw in Chapter 4) are more important than others structurally, and we organize our learning around them. This sort of plain old memorization is what musicians do when they learn the muscle movements necessary to play a particular piece; it is part of the reason that musicians like Kenny can't start playing on just any note, but tend to go to the beginnings of meaningful units, the beginnings of their hierarchically organized chunks.

* * *

Being an expert musician thus take many forms: dexterity at playing an instrument, emotional communication, creativity, and special mental structures for remembering music. Being an expert listener, which most of us are by age six, involves having incorporated the grammar of our musical culture into mental schemas that allow us to form musical expectations, the heart of the aesthetic experience in music. How all these various forms of expertise are acquired is still a neuroscientific mystery. The emerging consensus, however, is that musical expertise is not one thing, but involves many components, and not all musical experts will be endowed with these different components equally—some, like Irving Berlin, may lack what most of us would even consider a fundamental aspect of musicianship, being able to play an instrument well. It seems unlikely from what we now know that musical expertise is wholly different from expertise in other domains. Although music certainly uses brain structures and neural circuits that other activities don't, the process of becoming a musical expert—whether a composer or performer—requires many of the same personality traits as becoming an expert in other domains, especially diligence, patience, motivation, and plain old-fashioned stick-to-it-iveness.

Becoming a famous musician is another matter entirely, and may not have as much to do with intrinsic factors or ability as with charisma, opportunity, and luck. An essential point bears repeating, however: All of us are expert musical listeners, able to make quite subtle determinations of what we like and don't like, even when we're unable to articulate the reasons why. Science does have something to say about why we like the music we do, and that story is another interesting facet of the interplay between neurons and notes.

8. My Favorite Things

Why Do We Like the Music We Like?

You wake from a deep sleep and open your eyes. It's dark. The distant regular beating at the periphery of your hearing is still there. You rub your eyes with your hands, but you can't make out any shapes or forms. Time passes, but how long? Half an hour? One hour? Then you hear a different but recognizable sound—an amorphous, moving, wiggly sound with fast beating, a pounding that you can feel in your feet. The sounds start and stop without definition. Gradually building up and dying down, they weave together with no clear beginnings or endings. These familiar sounds are comforting, you've heard them before. As you listen, you have a vague notion of what will come next, and it does, even as the sounds remain remote and muddled, as though you're listening underwater.

Inside the womb, surrounded by amniotic fluid, the fetus hears sounds. It hears the heartbeat of its mother, at times speeding up, at other times slowing down. And the fetus hears music, as was recently discovered by Alexandra Lamont of Keele University in the UK. She found that, a year after they are born, children recognize and prefer music they were exposed to in the womb. The auditory system of the fetus is fully functional about twenty weeks after conception. In Lamont's experiment, mothers played a single piece of music to their babies repeatedly during the final three months of gestation. Of course, the babies were also hearing—through the waterlike filtering of the amniotic fluid in the womb—all of the sounds of their mothers' daily life, including other music, conversations, and environmental noises. But one particular piece was singled out for each baby to hear on a regular basis. The singled-out pieces included classical (Mozart, Vivaldi), Top 40 (Five, Backstreet Boys), reggae (UB40, Ken Boothe) and world beat (Spirits of Nature). After birth, the mothers were not allowed to play the experimental song to their infants. Then, one year later, Lamont played babies the music that they had heard in the womb, along with another piece of music chosen to be matched for style and tempo. For example, a baby who had heard UB40's reggae track "Many Rivers to Cross" heard that piece again, a year later, along with "Stop Loving You" by the reggae artist Freddie McGregor. Lamont then determined which one the babies preferred.

How do you know which of two stimuli a preverbal infant prefers? Most infant researchers use a technique known as the conditioned head-turning procedure, developed by Robert Fantz in the 1960s, and refined by John Columbo, Anne Fernald, the late Peter Jusczyk, and their colleagues. Two loudspeakers are set up in the laboratory and the infant is placed (usually on his mother's lap) between the speakers. When the infant looks at one speaker, it starts to play music or some other sound, and when he looks at the other speaker, it starts to play different music or a different sound. The infant quickly learns that he can control what is playing by where he is looking; he learns, that is, that the conditions of the experiment are under his control. The experimenters make sure that they counterbalance (randomize) the location that the different stimuli come from; that is, half the time the stimulus under study comes from one speaker and half the time it comes from the other. When Lamont did this with the infants in her study, she found that they tended to look longer at the speaker that was playing music they had heard in the womb than at the speaker playing the novel music, confirming that they preferred the music to which they had the prenatal exposure. A control group of one-year-olds who had not heard any of the music before showed no preference, confirming that there was nothing about the music itself that caused these results. Lamont also found that, all things being equal, the young infant prefers fast, upbeat music to slow music.

These findings contradict the long-standing notion of childhood amnesia—that we can't have any veridical memories before around the age of five. Many people claim to have memories from early childhood around age two and three, but it is difficult to know whether these are true memories of the original event, or rather, memory of someone telling us about the event later. The young child's brain is still undeveloped, functional specialization of the brain isn't complete, and neural pathways are still in the process of being made. The child's mind is trying to assimilate as much information as possible in as short a time as possible; there are typically large gaps in the child's understanding, awareness, or memory for events because he hasn't yet learned how to distinguish important events from unimportant ones, or to encode experience systematically. Thus, the young child is a prime candidate for suggestion, and could unwittingly encode, as his own, stories that were told to him about himself. It appears that for music even prenatal experience is encoded in memory, and can be accessed in the absence of language or explicit awareness of the memory.

A study made the newspapers and morning talk shows several years ago, claiming that listening to Mozart for ten minutes a day made you smarter ("the Mozart Effect"). Specifically, music listening, it was claimed, can improve your performance on spatial-reasoning tasks given immediately after the listening session (which some journalists thought implied mathematical ability as well). U.S. congressmen were passing resolutions, the governor of Georgia appropriated funds to buy a Mozart CD for every newborn baby Georgian. Most scientists found ourselves in an uncomfortable position. Although we do believe intuitively that music can enhance other cognitive skills, and although we would all like to see more governmental funding for school music programs, the actual study that claimed this contained many scientific flaws. The study was claiming some of the right things but for the wrong reasons. Personally, I found all the hubbub a bit offensive because the implication was that music should not be studied in and of itself, or for its own right, but only if it could help people to do better on other, "more important" things. Think how absurd this would sound if we turned it inside out. If I claimed that studying mathematics helped musical ability, would policy makers start pumping money into math for that reason? Music has often been the poor stepchild of public schools, the first program to get cut when there are funding problems, and people frequently try to justify it in terms of its collateral benefits, rather than letting music exist for its own rewards.

The problem with the "music makes you smarter" study turned out to be straightforward: The experimental controls were inadequate, and the tiny difference in spatial ability between the two groups, according to research by Bill Thompson, Glenn Schellenberg, and others, all turned on the choice of a control task. Compared to sitting in a room and doing nothing, music listening looked pretty good. But if subjects in the control task were given the slightest mental stimulation—hearing a book on tape, reading, etc.—there was no advantage for music listening. Another problem with the study was that there was no plausible mechanism proposed by which this might work—how could music listening increase spatial performance?

Glenn Schellenberg has pointed out the importance of distinguishing short-term from long-term effects of music. The Mozart Effect referred to immediate benefits, but other research *has* revealed long-term effects of musical activity. Music listening enhances or changes certain neural circuits, including the density of dendritic connections in the primary auditory cortex. The Harvard neuroscientist Gottfried Schlaug has shown that the front portion of the corpus callosum—the mass of fibers connecting the two cerebral hemispheres—is significantly larger in musicians than nonmusicians, and particularly for musicians who began their training early. This reinforces the notion that musical operations become bilateral with increased training, as musicians coordinate and recruit neural structures in both the left and right hemispheres.

Several studies have found microstructural changes in the cerebellum after the acquisition of motor skills, such as are acquired by musicians, including an increased number and density of synapses. Schlaug found that musicians tended to have larger cerebellums than nonmusicians, and an increased concentration of gray matter; gray matter is that part of the brain that contains the cell bodies, axons, and dendrites, and is understood to be responsible for information processing, as opposed to white matter, which is responsible for information transmission.

Whether these structural changes in the brain translate to enhanced abilities in nonmusical domains has not been proven, but music listening and music therapy have been shown to help people overcome a broad range of psychological and physical problems. But, to return to a more fruitful line of inquiry regarding musical taste . . . Lamont's results are important because they show that the prenatal and newborn brain are able to store memories and retrieve them over long periods of time. More practically, the results indicate that the environment—even when mediated by amniotic fluid and by the womb—can affect a child's development and preferences. So the seeds of musical preference are sown in the womb, but there must be more to the story than that, or children would simply gravitate toward the music their mothers like, or that plays in Lamaze classes. What we can say is that musical preferences are influenced, but not determined, by what we hear in the womb. There also is an extended period of acculturation, during which the infant takes in the music of the culture she is born into. There were reports a few years ago that prior to becoming used to the music of a foreign (to us) culture, all infants prefer Western music to other musics, regardless of their culture or race. These findings were not corroborated, but rather, it was found that infants do show a preference for consonance over dissonance. Appreciating dissonance comes later in life, and people differ in how much dissonance they can tolerate.

There is probably a neural basis for this. Consonant intervals and dissonant intervals are processed via separate mechanisms in the auditory cortex. Recent results from studying the electrophysiological responses of humans and monkeys to sensory dissonance (that is, chords that sound dissonant by virtue of their frequency ratios, not due to any harmonic or musical context) show that neurons in the primary auditory cortex—the first level of cortical processing for sound—synchronize their firing rates during dissonant chords, but not during consonant chords. Why that would create a preference for consonance is not yet clear.

We do know a bit about the infant's auditory world. Although infant ears are fully functioning four months before birth, the developing brain requires months or years to reach full auditory processing capacity. Infants recognize transpositions of pitch and of time (tempo changes), indicating they are capable of relational processing, something that even the most advanced computers still can't do very well. Jenny Saffran of the University of Wisconsin and Laurel Trainor of McMaster University have gathered evidence that infants can also attend to absolute-pitch cues if the task requires it, suggesting a cognitive flexibility previously unknown: Infants can employ different modes of processing—presumably mediated by different neural circuits—depending on what will best help them to solve the problem at hand.

Trehub, Dowling, and others have shown that contour is the most salient musical feature for infants, who can detect contour similarities and differences even across thirty seconds of retention. Recall that *contour* refers to the pattern of musical pitch in a melody—the sequence of ups and downs that the melody takes—regardless of the size of the interval. Someone attending to contour exclusively would encode only that the melody goes up, for example, but not by how much. Infants' sensitivity to musical contour parallels their sensitivity to linguistic contours—which separate questions from exclamations, for example, and which are part of what linguists call prosody. Fernald and Trehub have documented the ways in which parents speak differently to infants than to older children and adults, and this holds across cultures. The resulting manner of speaking uses a slower tempo, an extended pitch range, and a higher overall pitch level.

Mothers (and to a lesser extent, fathers) do this quite naturally without any explicit instruction to do so, using an exaggerated intonation that the researchers call infant-directed speech or motherese. We believe that motherese helps to call the babies' attention to the mother's voice, and helps to distinguish words within the sentence. Instead of saying, as we would to an adult, "This is a ball," motherese would entail something like, "Seeeeee?" (with the pitch of the eee's going up to the end of the sentence). "See the BAAAAAALLLLLL?" (with the pitch covering an extended range and going up again at the end of the word *ball*). In such utterances, the contour is a signal that the mother is asking a question or making a statement, and by exaggerating the differences between up and down contours, the mother calls attention to them. In effect, the mother is creating a prototype for a question and a prototype for a declaration, and ensuring that the prototypes are easily distinguishable. When a mother gives an exclamatory scold, quite naturally—and again without explicit training—she is likely to create a third type of prototypical utterance, one that is short and clipped, without much pitch variation: "No!" (pause) "No! Bad!" (pause) "I said no!" Babies seem to come hardwired with an ability to detect and track contour, preferentially, over specific pitch intervals.

Trehub also showed that infants are more able to encode consonant intervals such as perfect fourth and perfect fifth than dissonant ones, like the tritone. Trehub found that the unequal steps of our scale make it easier to process intervals even early in infancy. She and her colleagues played nine-month-olds the regular seven-note major scale and two scales she invented. For one of these invented scales, she divided the octave into eleven equal-space steps and then selected seven tones that made one- and two-step patterns, and for the other she divided the octave into seven equal steps. The infants' task was to detect a mistuned tone. Adults performed well with the major scale, but poorly with both of the artificial, never-before-heard scales. In contrast, the infants did equally well on both unequally tuned scales and on the equally tuned ones. From prior work, it is believed that nine-month-olds have not yet incorporated a mental schema for the major scale, so this suggests a general processing advantage for unequal steps, something our major scale has.

In other words, our brains and the musical scales we use seem to have coevolved. It is no accident that we have the funny, asymmetric arrangement of notes in the major scale: It is easier to learn melodies with this arrangement, which is a result of the physics of sound production (via the overtone series we visited earlier); the set of tones we use in our major scale are very close in pitch to the tones that constitute the overtone series. Very early in childhood, most children start to spontaneously vocalize, and these early vocalizations can sound a lot like singing. Babies explore the range of their voices, and begin to explore phonetic production, in response to the sounds they are bringing in from the world around them. The more music they hear, the more likely they are to include pitch and rhythmic variations in their spontaneous vocalizations.

Young children start to show a preference for the music of their culture by age two, around the same time they begin to develop specialized speech processing. At first, children tend to like simple songs, where *simple* means music that has clearly defined themes (as opposed to, say, four-part counterpoint) and chord progressions that resolve in direct and easily predictable ways. As they mature, children start to tire of easily predictable music and search for music that holds more challenge. According to Mike Posner, the frontal lobes and the anterior cingulate—a structure just behind the frontal lobes that directs attention—are not fully formed in children, leading to an inability to pay attention to several things at once; children show difficulty attending to one stimulus when distracters are present. This accounts for why children under the age of eight or so have so much difficulty singing "rounds" like "Row, Row, Row Your Boat." Their attentional system—specifically the network that connects the cingulate gyrus (the larger structure within which the anterior cingulate sits) and the orbitofrontal regions of the brain—cannot adequately filter out unwanted or distracting stimuli. Children who have not yet reached the developmental stage of being able to exclude irrelevant auditory information face a world of great sonic complexity with all sounds coming in as a sensory barrage. They may try to follow the part of the song that their group is supposed to be singing, only to be distracted and tripped up by the competing parts in the round. Posner has shown that certain exercises adapted from attention and concentration games used by NASA can help accelerate the development of the child's attentional ability.

The developmental trajectory, in children, of first preferring simple and then more complex songs is a generalization, of course; not all children like music in the first place, and some children develop a taste for music that is off the beaten path, oftentimes through pure serendipity. I became fascinated with big band and swing music when I was eight, around the time my grandfather gave me his collection of 78 rpm records from the World War II era. I was initially attracted by novelty songs, such as "The Syncopated Clock," "Would You Like to Swing on a Star," "The Teddy Bear's Picnic," and "Bibbidy Bobbidy Boo"—songs that were made for children. But sufficient exposure to the relatively exotic chord patterns and voicings of Frank de Vol's and Leroy Anderson's orchestras became part of my mental wiring, and I soon found myself listening to all kinds of jazz; the children's jazz opened the neural doors to make jazz in general palatable and understandable.

Researchers point to the teen years as the turning point for musical preferences. It is around the age of ten or eleven that most children take on music as a real interest, even those children who didn't express such an interest in music earlier. As adults, the music we tend to be nostalgic for, the music that feels like it is "our" music, corresponds to the music we heard during these years. One of the first signs of Alzheimer's disease (a disease characterized by changes in nerve cells and neurotransmitter levels, as well as destruction of synapses) in older adults is memory loss. As the disease progresses, memory loss becomes more profound. Yet many of these old-timers can still remember how to sing the songs they heard when they were fourteen. Why fourteen? Part of the reason we remember songs from our teenage years is because those years were times of self-discovery, and as a consequence, they were emotionally charged; in general, we tend to remember things that have an emotional component because our amygdala and neurotransmitters act in concert to "tag" the memories as something important. Part of the reason also has to do with neural maturation and pruning; it is around fourteen that the wiring of our musical brains is approaching adultlike levels of completion.

There doesn't seem to be a cutoff point for acquiring new tastes in music, but most people have formed their tastes by the age of eighteen or twenty. Why this is so is not clear, but several studies have found it to be the case. Part of the reason may be that in general, people tend to become less open to new experiences as they age. During our teenage years, we begin to discover that there exists a world of different ideas, different cultures, different people. We experiment with the idea that we don't have to limit our life's course, our personalities, or our decisions to what we were taught by our parents, or to the way we were brought up. We also seek out different kinds of music. In Western culture in particular, the choice of music has important social consequences. We listen to the music that our friends listen to. Particularly when we are young, and in search of our identity, we form bonds or social groups with people whom we want to be like, or whom we believe we have something in common with. As a way of externalizing the bond, we dress alike, share activities, and listen to the same music. Our group listens to this kind of music, those people listen to that kind of music. This ties into the evolutionary idea of music as a vehicle for social bonding and societal cohesion. Music and musical preferences become a mark of personal and group identity and of distinction.

To some degree, we might say that personality characteristics are associated with, or predictive of, the kind of music that people like. But to a large degree, it is determined by more or less chance factors: where you went to school, who you hung out with, what music they happened to be listening to. When I lived in northern California as a kid, Creedence Clearwater Revival was huge—they were from just down the road. When I moved to southern California, CCR's brand of quasi-cowboy, country-hick music didn't fit in well with the surfer/Hollywood culture that embraced the Beach Boys and more theatrical performance artists like David Bowie.

Also, our brains are developing and forming new connections at an explosive rate throughout adolescence, but this slows down substantially after our teenage years, the formative phase when our neural circuits become structured out of our experiences. This process applies to the music we hear; new music becomes assimilated within the framework of the music we were listening to during this critical period. We know that there are critical periods for acquiring new skills, such as language. If a child doesn't learn language by the age of six or so (whether a first or a second language), the child will never learn to speak with the effortlessness that characterizes most native speakers of a language. Music and mathematics have an extended window, but not an unlimited one: If a student hasn't had music lessons or mathematical training prior to about age twenty, he can still learn these subjects, but only with great difficulty, and it's likely that he will never "speak" math or music like someone who learned them early. This is because of the biological course for synaptic growth. The brain's synapses are programmed to grow for a number of years, making new connections. After that time, there is a shift toward pruning, to get rid of unneeded connections.

Neuroplasticity is the ability of the brain to reorganize itself. Although in the last five years there have been some impressive demonstrations of brain reorganization that used to be thought impossible, the amount of reorganization that can occur in most adults is vastly less than can occur in children and adolescents.

Of course, there are individual differences. Just as some people can heal broken bones or skin cuts faster than others, so, too, can some people forge new connections more easily than others. Generally, between the ages of eight and fourteen, pruning starts to occur in the frontal lobes, the seat of higher thought and reasoning, planning, and impulse control. Myelination starts to ramp up during this time. Myelin is a fatty substance that coats the axons, speeding up synaptic transmission. (This is why as children get older, generally, problem solving becomes more rapid and they are able to solve more complex problems.) Myelination of the whole brain is generally completed by age twenty. Multiple sclerosis is one of several degenerative diseases that can affect the myelin sheath surrounding the neurons.

The balance between simplicity and complexity in music also informs our preferences. Scientific studies of like and dislike across a variety of aesthetic domains—painting, poetry, dance, and music—have shown that an orderly relationship exists between the complexity of an artistic work and how much we like it. Of course, complexity is an entirely subjective concept. In order for the notion to make any sense, we have to allow for the idea that what seems impenetrably complex to Stanley might fall right in the "sweet spot" of preference for Oliver. Similarly, what one person finds insipid and hideously simple, another person might find difficult to understand, based on differences in background, experience, understanding, and cognitive schemas.

In a sense, schemas are everything. They frame our understanding; they're the system into which we place the elements and interpretations of an aesthetic object. Schemas inform our cognitive models and expectations. With one schema, Mahler's Fifth is perfectly interpretable, even upon hearing it for the first time: It is a symphony, it follows symphonic form with four movements; it contains a main theme and subthemes, and repetitions of the theme; the themes are manifested through orchestral instruments, as opposed to African talking drums or fuzz bass. Those familiar with Mahler's Fourth will recognize that the Fifth opens with a variation on that same theme, and even at the same pitch. Those well acquainted with Mahler's work will recognize that the composer includes quotations from three of his own songs. Musically educated listeners will be aware that most symphonies from Haydn to Brahms and Bruckner typically begin and end in the same key. Mahler flouts this convention with his Fifth, moving from C-sharp minor to A minor and finally ending in D major. If you had not learned to hold in your mind a sense of key as the symphony develops, or if you did not have a sense of the normal trajectory of a symphony, this would be meaningless; but for the seasoned listener, this flouting of convention brings a rewarding surprise, a violation of expectations, especially when such key changes are done skillfully so as not to be jarring. Lacking a proper symphonic schema, or if the listener holds another schema, perhaps that of an aficionado of Indian ragas, Mahler's Fifth is nonsensical or perhaps rambling, one musical idea melding amorphously into the next, with no boundaries, no beginnings or endings that appear as part of a coherent whole. The schema frames our perception, our cognitive processing, and ultimately our experience.

When a musical piece is too simple we tend not to like it, finding it trivial. When it is too complex, we tend not to like it, finding it unpredictable—we don't perceive it to be grounded in anything familiar. Music, or any art form for that matter, has to strike the right balance between simplicity and complexity in order for us to like it. Simplicity and complexity relate to familiarity, and *familiarity* is just another word for a schema.

It is important in science, of course, to define our terms. What is "too simple" or "too complex"? An operational definition is that we find a piece too simple when we find it trivially predictable, similar to something we have experienced before, and without the slightest challenge. By analogy, consider the game tic-tac-toe. Young children find it endlessly fascinating, because it has many features that contribute to interest at their level of cognitive ability: It has clearly defined rules that any child can easily articulate; it has an element of surprise in that the player never knows for sure exactly what her opponent will do next; the game is dynamic, in that one's own next move is influenced by what one's opponent did; when the game will end, who will win, or whether it will be a draw is undetermined, yet there is an outer limit of nine moves. That indeterminacy leads to tension and expectations, and the tension is finally released when the game is over.

As the child develops increasing cognitive sophistication, she eventually learns strategies—the person who moves second cannot win against a competent player; the best the second player can hope for is a draw. When the sequence of moves and the end point of the game become predictable, tic-tac-toe loses its appeal. Of course, adults can still enjoy playing the game with children, but we enjoy seeing the pleasure on the child's face and we enjoy the process—spread out over several years—of the child learning to unlock the mysteries of the game as her brain develops.

To many adults, Raffi and Barney the Dinosaur are the musical equivalents of tic-tac-toe. When music is too predictable, the outcome too certain, and the "move" from one note or chord to the next contains no element of surprise, we find the music unchallenging and simplistic. As the music is playing (particularly if you're engaged with focused attention), your brain is thinking ahead to what the different possibilities for the next note are, where the music is going, its trajectory, its intended direction, and its ultimate end point. The composer has to lull us into a state of trust and security; we have to allow him to take us on a harmonic journey; he has to give us enough little rewards—completions of expectations—that we feel a sense of order and a sense of place.

Say you're hitchhiking from Davis, California, to San Francisco. You want the person who picks you up to take the normal route, Highway 80. You might be willing to tolerate a few shortcuts, especially if the driver is friendly, believable, and is up-front about what he's doing. ("I'm just going to cut over here on Zamora Road to avoid some construction on the freeway.") But if the driver takes you out on back roads with no explanation, and you reach a point where you no longer see any landmarks, your sense of safety is sure to be violated. Of course, different people, with different personality types, react differently to such unanticipated journeys, musical or vehicular. Some react with sheer panic ("That Stravinsky is going to kill me!") and some react with a sense of adventure at the thrill of discovery ("Coltrane is doing something weird here, but what the hell, it won't hurt me to stick around awhile longer, I can take care of my harmonic self and find my way back to musical reality if I have to").

To continue the analogy with games, some games have such a complicated set of rules that the average person doesn't have the patience to learn them. The possibilities for what can happen on any given turn are too numerous or too unpredictable (to the novice) to contemplate. But an inability to predict what will happen next is not always a sign that a game holds eventual interest if only one sticks with it long enough. A game may have a completely unpredictable course no matter how much practice you have with it—many board games simply involve rolling the dice and waiting to see what happens to you. Chutes and Ladders and Candy Land are like this. Children enjoy the sense of surprise, but adults can find the game tedious because, although no one can predict exactly what will happen (the game is a function of the random throw of the dice), the outcome has no structure whatsoever, and moreover, there is no amount of skill on the part of the player that can influence the course of the game.

Music that involves too many chord changes, or unfamiliar structure, can lead many listeners straight to the nearest exit, or to the "skip" button on their music players. Some games, such as Go, Axiom, or Zendo are sufficiently complicated or opaque to the novice that many people give up before getting very far: The structure presents a steep learning curve, and the novice can't be sure that the time invested will be worth it. Many of us have the same experience with unfamiliar music, or unfamiliar musical forms. People may tell you that Schönberg is brilliant, or that Tricky is the next Prince, but if you can't figure out what is going on in the first minute or so of one of their pieces, you may find yourself wondering if the payoff will justify the effort you spend trying to sort it all out. We tell ourselves that if we only listen to it enough times, we may begin to understand it and to like it as much as our friends do. Yet, we recall other times in our lives when we invested hours of prime listening time in an artist and never arrived at the point where we "got it." Trying to appreciate new music can be like contemplating a new friendship in that it takes time, and sometimes there is nothing you can do to speed it up. At a neural level, we need to be able to find a few landmarks in order to invoke a cognitive schema. If we hear a piece of radically new music enough times, some of that piece will eventually become encoded in our brains and we will develop landmarks. If the composer is skillful, those parts of the piece that become our landmarks will be the very ones that the composer intended they should be; his knowledge of composition and human perception and memory will have allowed him to create certain "hooks" in the music that will eventually stand out in our minds.

Structural processing is one source of difficulty in appreciating a new piece of music. Not understanding symphonic form, or the sonata form, or the AABA structure of a jazz standard, is the music-listening equivalent of driving on a highway with no road signs: You never know where you are or when you'll arrive at your destination (or even at an interim spot that is not your destination, but one that provides an orienting landmark). For example, many people just don't "get" jazz; they say that it sounds like an unstructured, crazy, and formless improvisation, a musical competition to squeeze as many notes as possible into as small a space as possible. There are more than a half-dozen subgenres of what people collectively call "jazz": Dixieland, boogie-woogie, big band, swing, bebop, "straight-ahead," acid-jazz, fusion, metaphysical, and so on. "Straight-ahead," or "classic jazz," as it is sometimes called, is more or less the standard form of jazz, analogous to the sonata or the symphony in classical music, or what a typical song by the Beatles or Billy Joel or the Temptations is to rock music.

In classic jazz, the artist begins by playing the main theme of the song; often a well-known one from Broadway, or one that has already been a hit for someone else; such songs are called "standards," and they include "As Time Goes By," "My Funny Valentine," and "All of Me." The artist runs through the complete form of the song once—typically two verses and the chorus (otherwise known as a "refrain"), followed by another verse. The chorus is the part of a song that repeats regularly throughout; the verses are what change. We call this form AABA, where the letter *A* represents the verse and the letter *B* represents the chorus. AABA means we play verse-verse-chorus-verse. Many other variations are possible, of course. Some songs have a C section, called the bridge.

The term *chorus* is used to mean not just the second section of the song, but also one run through the entire form. In other words, running through the AABA portion of a song once is called "playing one chorus."

When I play jazz, if someone says, "Play the chorus," or, "Let's go over the chorus" (using the word *the*), we all assume he means a section of the song. If, instead someone says, "Let's run through one chorus," or, "Let's do a couple of choruses," we know he means the entire form.

"Blue Moon" (Frank Sinatra, Billie Holiday) is an example of a song with AABA form. A jazz artist may play around with the rhythm or feel of the song, and may embellish the melody. After playing through the form of the song once, what jazz musicians refer to as "the head," the different members of the ensemble take turns improvising new music over the chord progression and form of the original song. Each musician plays through one or more choruses and then the next musician takes over at the beginning of the head. During the improvisations, some artists stick close to the original melody, some add ever distant and exotic harmonic departures. When everyone has had a chance to improvise, the band returns to the head, playing it more or less straight, and then they're done. The improvisations can go on for many minutes—it is not uncommon for a jazz rendition of a two- or three-minute song to stretch out to ten to fifteen minutes. There is also a typical order to how the musicians take turns: The horns go first, followed by the piano and/or guitar, followed by the bass player. Sometimes the drummer also improvises, and he would typically follow the bass. Sometimes the musicians also share part of a chorus—each musician playing four or eight measures, and then handing off the solo to another musician, a sort of musical relay race.

To the newbie, the whole thing may seem chaotic. Yet, simply knowing that the improvisation takes place over the original chords and form of the song can make a big difference in orienting the neophyte to where in the song the players are. I often advise new listeners to jazz to simply hum the main tune in their mind once the improvisation begins—this is what the improvisers themselves are often doing—and that enriches the experience considerably.

Each musical genre has its own set of rules and its own form. The more we listen, the more those rules become instantiated in memory. Unfamiliarity with the structure can lead to frustration or a simple lack of appreciation. Knowing a genre or style is to effectively have a category built around it, and to be able to categorize new songs as being either members or nonmembers of that category—or in some cases, as "partial" or "fuzzy" members of the category, members subject to certain exceptions.

The orderly relationship between complexity and liking is referred to as the inverted-U function because of the way a graph would be drawn that relates these two factors. Imagine a graph in which the x-axis is how complex a piece of music is (to you) and the y-axis is how much you like it. At the bottom left of this graph, close to the origin, there would be a point for music that is very simple and your reaction being that you don't like it. As the music increases in complexity, your liking increases as well. The two variables follow each other for quite a while on the graph—increased complexity yields increased liking, until you cross some personal threshold and go from disliking the piece intensely to actually liking it quite a bit. But at some point as we increase complexity, the music becomes too complex, and your liking for it begins to decrease. Now more complexity in the music leads to less and less liking, until you cross another threshold and you no longer like the music at all. Too complex and you absolutely hate the music. The shape of such a graph would make an inverted U or an inverted V.

The inverted-U hypothesis is not meant to imply that the only reason you might like or dislike a piece of music is because of its simplicity or complexity. Rather, it is intended to account for this variable. The elements of music can themselves form a barrier to appreciation of a new piece of music. Obviously, if music is too loud or too soft, this can be problematic. But even the dynamic range of a piece—the disparity between the loudest and softest parts—can cause some people to reject it. This can be especially true for people who use music to regulate their mood in a specific way. Someone who wants music to calm her down, or someone else who wants music to pep him up for a workout, is probably not going to want to hear a musical piece that runs the loudness gamut all the way from very soft to very loud, or emotionally from sad to exhilarating (as does Mahler's Fifth, for example). The dynamic range as well as the emotional range is simply too wide, and may create a barrier to entry.

Pitch can also play into preference. Some people can't stand the thumping low beats of modern hip-hop, others can't stand what they describe as the high-pitched whininess of violins. Part of this may be a matter of physiology; literally, different ears may transmit different parts of the frequency spectrum, causing some sounds to appear pleasant and others aversive. There may also exist psychological associations, both positive and negative, to various instruments.

Rhythm and rhythmic patterns influence our ability to appreciate a given musical genre or piece. Many musicians are drawn to Latin music because of the complexity of the rhythms. To an outsider, it all just sounds "Latin," but to someone who can make out the nuances of when a certain beat is strong relative to other beats, Latin music is a whole world of interesting complexity: bossa nova, samba, rhumba, beguine, mambo, merengue, tango—each is a completely distinct and identifiable style of music. Some people genuinely enjoy Latin music and Latin rhythms without being able to tell them apart, of course, but others find the rhythms too complicated and unpredictable, and this is a turnoff to them. I've found that if I teach one or two Latin rhythms to listeners, they come to appreciate them; it is all a question of grounding and having a schema. For other listeners, rhythms that are too simple are the dealbreaker for a style of music. The typical complaint of my parents' generation about rock and roll, apart from how loud it seemed to them, was that it all had the same beat.

Timbre is another barrier for many people and its influence is almost certainly increasing, as I argued in Chapter 1. The first time I heard John Lennon or Donald Fagen sing, I thought the voices unimaginably strange. I didn't want to like them. Something kept me going back to listen, though—perhaps it was the strangeness—and they wound up being two of my favorite voices; voices that now have gone beyond familiar and approach what I can only call intimate; I feel as though these voices have become incorporated into who I am. And at a neural level, they have. Having listened to thousands of hours of both these singers, and tens of thousands of playings of their songs, my brain has developed circuitry that can pick out their voices from among thousands of others, even when they sing something I've never heard them sing before. My brain has encoded every vocal nuance and every timbral flourish, so that if I hear an alternate version of one of their songs—as we do on the *John Lennon Collection* of demo versions of his albums—I can immediately recognize the ways in which this performance deviates from the one I have stored in the neural pathways of my long-term memory.

As with other sorts of preferences, our musical preferences are also influenced by what we've experienced before, and whether the outcome of that experience was positive or negative. If you had a negative experience once with pumpkin—say, for example, it made you sick to your stomach—you are likely to be wary of future excursions into pumpkin gustation. If you've had only a few, but largely positive, encounters with broccoli, you might be willing to try a new broccoli recipe, perhaps broccoli soup, even if you've never had it before. The one positive experience begets others.

The types of sounds, rhythms, and musical textures we find pleasing are generally extensions of previous positive experiences we've had with music in our lives. This is because hearing a song that you like is a lot like having any other pleasant sensory experience—eating chocolate, fresh-picked raspberries, smelling coffee in the morning, seeing a work of art or the peaceful face of someone you love who is sleeping. We take pleasure in the sensory experience, and find comfort in its familiarity and the safety that familiarity brings. I can look at a ripe raspberry, smell it, and anticipate that it will taste good and that the experience will be safe—I won't get sick. If I've never seen a loganberry before, there are enough points in common with the raspberry that I can take the chance in eating it and anticipate that it will be safe.

Safety plays a role for a lot of us in choosing music. To a certain extent, we surrender to music when we listen to it—we allow ourselves to trust the composers and musicians with a part of our hearts and our spirits; we let the music take us somewhere outside of ourselves. Many of us

feel that great music connects us to something larger than our own existence, to other people, or to God. Even when music doesn't transport us
to an emotional place that is transcendent, music can change our mood.
We might be understandably reluctant, then, to let down our guard, to
drop our emotional defenses, for just anyone. We will do so if the musicians and composer make us feel safe. We want to know that our vulnerability is not going to be exploited. This is part of the reason why so
many people can't listen to Wagner. Due to his pernicious anti-Semitism,
the sheer vulgarity of his mind (as Oliver Sacks describes it), and his
music's association with the Nazi regime, some people don't feel safe listening to his music. Wagner has always disturbed me profoundly, and not
just his music, but also the *idea* of listening to it. I feel reluctant to give
into the seduction of music created by so disturbed a mind and so dangerous (or impenetrably hard) a heart as his, for fear that I might develop
some of the same ugly thoughts. When I listen to the music of a great
composer I feel that I am, in some sense, becoming one with him, or letting a part of him inside me. I also find this disturbing with popular music, because surely some of the purveyors of pop are crude, sexist, racist,
or all three.

This sense of vulnerability and surrender is no more prevalent than
with rock and popular music in the past forty years. This accounts for
the fandom that surrounds popular musicians—the Grateful Dead, the
Dave Matthews Band, Phish, Neil Young, Joni Mitchell, the Beatles,
R.E.M., Ani DiFranco. We allow them to control our emotions and even
our politics—to lift us up, to bring us down, to comfort us, to inspire us.
We let them into our living rooms and bedrooms when no one else is
around. We let them into our ears, directly, through earbuds and headphones, when we're not communicating with anybody else in the world.

It is unusual to let oneself become so vulnerable with a total stranger.
Most of us have some kind of protection that prevents us from blurting
out every thought and feeling that comes across our minds. When someone asks us, "How're ya doin'?" we say, "Fine," even if we're depressed
about a fight we just had at home, or suffering a minor physical ailment.
My grandfather used to say that the definition of a bore is someone who

when you ask him "How are you?" actually tells you. Even with close friends, there are some things we simply keep hidden—digestive and bowel-related problems, for example, or feelings of self-doubt. One of the reasons that we're willing to make ourselves vulnerable to our favorite musicians is that they often make themselves vulnerable to us (or they convey vulnerability through their art—the distinction between whether they are actually vulnerable or merely representing it artistically is not important for now).

The power of art is that it can connect us to one another, and to larger truths about what it means to be alive and what it means to be human. When Neil Young sings

Old man look at my life, I'm a lot like you were. . . .
Live alone in a paradise that makes me think of two.

we feel for the man who wrote the song. I may not live in a paradise, but I can empathize with a man who may have some material success but no one to share it with, a man who feels he has "gained the world but lost his soul," as George Harrison once sang, quoting at once the gospel according to Mark and Mahatma Gandhi.

Or when Bruce Springsteen sings "Back in Your Arms" about losing love, we resonate to a similar theme, by a poet with a similar "everyman" persona to Neil Young's. And when we consider how much Springsteen has—the adoration of millions of people worldwide, and millions of dollars—it becomes all the more tragic that he cannot have the one woman he wants.

We hear vulnerability in unlikely places and it brings us closer to the artist. David Byrne (of the Talking Heads) is generally known for his abstract, arty lyrics, with a touch of the cerebral. In his solo performance of "Lilies of the Valley," he sings about being alone and scared. Part of our appreciation for this lyric is enhanced by knowing something about the artist, or at least the artist's persona, as an eccentric intellectual, who rarely revealed something as raw and transparent as being afraid.

Connections to the artist or what the artist stands for can thus be part of our musical preferences. Johnny Cash cultivated an outlaw image, and also showed his compassion for prison inmates by performing many concerts in prisons. Prisoners may like Johnny Cash's music—or grow to like it—because of what the artist stands for, quite apart from any strictly musical considerations. But fans will only go so far to follow their heroes, as Dylan learned at the Newport Folk Festival. Johnny Cash could sing about wanting to leave prison without alienating his audience, but if he had said that he liked visiting prisons because it helped him appreciate his own freedom, he would no doubt have crossed a line from compassion to gloating, and his inmate audience would have understandably turned on him.

Preferences begin with exposure and each of us has our own "adventuresomeness" quotient for how far out of our musical safety zone we are willing to go at any given time. Some of us are more open to experimentation than others in all aspects of our lives, including music; and at various times in our life we may seek or avoid experimentation. Generally, the times when we find ourselves bored are those when we seek new experiences. As Internet radio and personal music players are becoming more popular, I think that we will be seeing personalized music stations in the next few years, in which everyone can have his or her own personal radio station, controlled by computer algorithms that play us a mixture of music we already know and like and a mixture of music we don't know but we are likely to enjoy. I think it will be important that whatever form this technology takes, listeners should have an "adventuresomeness" knob they can turn that will control the mix of old and new, or the mix of how far out the new music is from what they usually listen to. This is something that is highly variable from person to person, and even, within one person, from one time of day to the next.

Our music listening creates schemas for musical genres and forms, even when we are only listening passively, and not attempting to analyze the music. By an early age, we know what the legal moves are in the music of our culture. For many, our future likes and dislikes will be a consequence of the types of cognitive schemas we formed for music through childhood listening. This isn't meant to imply that the music we listen to as children will necessarily determine our musical tastes for the rest of our lives; many people are exposed to or study music of different cultures and styles and become acculturated to them, learning their schemas as well. The point is that our early exposure is often our most profound, and becomes the foundation for further musical understanding.

Musical preferences also have a large social component based on our knowledge of the singer or musician, on our knowledge of what our family and friends like, and knowledge of what the music stands for. Historically, and particularly evolutionarily, music has been involved with social activities. This may explain why the most common form of musical expression, from the Psalms of David to Tin Pan Alley to contemporary music, is the love song, and why for most of us, love songs seem to be among our favorite things.

9. The Music Instinct

Evolution's #1 Hit

Where did music come from? The study of the evolutionary origins of music has a distinguished history, dating back to Darwin himself, who believed that it developed through natural selection as part of human or paleohuman mating rituals. I believe that the scientific evidence supports this idea as well, but not everyone agrees. After decades of only scattered work on the topic, in 1997 interest was suddenly focused on a challenge issued by the cognitive psychologist and cognitive scientist Steven Pinker.

There are about 250 people worldwide who study music perception and cognition as a primary research focus. As with most scientific disciplines, we hold conferences once a year. In 1997, the conference was held at MIT, and Steven Pinker was invited to give the opening address. Pinker had just completed *How the Mind Works*, an important large-scale work that explains and synthesizes the major principles of cognitive science, but he had not yet found popular notoriety. "Language is clearly an evolutionary adaptation," he told us during his keynote speech. "The cognitive mechanisms that we, as cognitive psychologists and cognitive scientists, study, mechanisms such as memory, attention, categorization, and decision making, all have a clear evolutionary purpose." He explained that, once in a while, we find a behavior or attribute in an organism that lacks any clear evolutionary basis; this occurs when evolutionary forces propagate an adaptation for a particular reason, and something else comes along for the ride, what Stephen Jay Gould called a spandrel, borrowing the term from architecture. In architecture, a designer might plan for a dome to be held up by four arches. There will necessarily be a space between the arches, not because it was planned for, but because it is a by-product of the design. Birds evolved feathers to keep warm, but they coopted the feathers for another purpose—flying. This is a spandrel.

Many spandrels are put to such good use that it is hard to know after the fact whether they were adaptations or not. The space between arches in a building became a place where artists painted angels and other decorations. The spandrel—a by-product of the architects' design—became one of the most beautiful parts of a building. Pinker argued that language is an adaptation and music is its spandrel. Among the cognitive operations that humans perform, music is the least interesting to study because it is merely a by-product, he went on, an evolutionary accident piggybacking on language.

"Music is auditory cheesecake," he said dismissively. "It just happens to tickle several important parts of the brain in a highly pleasurable way, as cheesecake tickles the palate." Humans didn't evolve a liking for cheesecake, but we did evolve a liking for fats and sugars, which were in short supply during our evolutionary history. Humans evolved a neural mechanism that caused our reward centers to fire when eating sugars and fats because in the small quantities they were available, they were beneficial to our well-being.

Most activities that are important for survival of the species, such as eating and sex, are also pleasurable; our brains evolved mechanisms to reward and encourage these behaviors. But we can learn to short-circuit the original activities and tap directly into these reward systems. We can eat foods that have no nutritive value and we can have sex without procreating; we can take heroin, which exploits the normal pleasure receptors in the brain; none of these is adaptive, but the pleasure centers in our limbic system don't know the difference. Humans, then, discovered that cheesecake just happens to push pleasure buttons for fat and sugar, Pinker explained, and music is simply a pleasure-seeking behavior that exploits one or more existing pleasure channels that evolved to reinforce an adaptive behavior, presumably linguistic communication.

"Music," Pinker lectured us, "pushes buttons for language ability (with which music overlaps in several ways); it pushes buttons in the auditory cortex, the system that responds to the emotional signals in a human voice crying or cooing, and the motor control system that injects rhythm into the muscles when walking or dancing."

"As far as biological cause and effect are concerned," Pinker wrote in *The Language Instinct* (and paraphrased in the talk he gave to us), "music is useless. It shows no signs of design for attaining a goal such as long life, grandchildren, or accurate perception and prediction of the world. Compared with language, vision, social reasoning, and physical know-how, music could vanish from our species and the rest of our lifestyle would be virtually unchanged."

When a brilliant and respected scientist such as Pinker makes a controversial claim, the scientific community takes notice, and it caused me and many of my colleagues to reevaluate a position on the evolutionary basis of music that we had taken for granted, without questioning. Pinker got us thinking. And a little research showed that he is not the only theorist to deride music's evolutionary origins. The cosmologist John Barrow said that music has no role in survival of the species, and psychologist Dan Sperber called music "an evolutionary parasite." Sperber believes that we evolved a cognitive capacity to process complex sound patterns that vary in pitch and duration, and that this communicative ability first arose in primitive, prelinguistic humans. Music, according to Sperber, developed parasitically to exploit this capacity that had evolved for true communication. Ian Cross of Cambridge University sums up: "For Pinker, Sperber, and Barrow, music exists simply because of the pleasure that it affords; its basis is purely hedonic."

I happen to think that Pinker is wrong, but I'll let the evidence speak for itself. Let me back up first a hundred and fifty years to Charles Darwin. The catchphrase most of us are taught in school, "survival of the fittest" (unfortunately propagated by the British philosopher Herbert Spencer), is an oversimplification of evolution. The theory of evolution rests on several assumptions. First, all of our phenotypic attributes (our appearance, physiological attributes, and some behaviors) are encoded in our genes, which are passed from one generation to the next. Genes tell our body how to make proteins, which generate our phenotypic characteristics. The action of genes is specific to the cells in which they reside; a given gene may contain information that is useful or not useful depending on the cell in question—cells in your eye don't need to grow skin, for example. Our genotype (particular sequence of DNA) gives rise to our phenotype (particular physical characteristics). So to sum up: Many of the ways in which members of a species differ from one another are encoded in the genes, and these are passed on through reproduction.

The second assumption of evolutionary theory is that there exists between us some natural genetic variability. Third, when we mate, our genetic material combines to form a new being, 50 percent of whose genetic material comes from each parent. Finally, due to spontaneous errors, mistakes or mutations sometimes occur that may be passed on to the next generation.

The genes that exist in you today (with the exception of a small number that may have mutated) are those that reproduced successfully in the past. Each of us is a victor in a genetic arms race; many genes that failed to reproduce successfully died out, leaving no descendants. Everyone alive today is composed of genes that won a long-lasting, large-scale genetic competition. "Survival of the fittest" is an oversimplification because it leads to the distorted view that genes that confer a survival advantage in their host organism are those that will win the genetic race. But living a long life, however happy and productive, does not pass on genes. An organism needs to reproduce to pass on its genes. The name of the evolutionary game is to reproduce at all costs, and to see that one's offspring live to do the same, and for their offspring to live long enough to do the same, and so on.

If an organism lives long enough to reproduce, and if its children are hearty and protected so that they can do the same, there is no compelling evolutionary reason for the organism to live a long time. Some avian species and spiders die during or after sexual mating. The post-mating years do not confer any advantage to the survival of the organism's genes unless it is able to use that time to protect its offspring, secure resources for them, or help them to find mates. Thus, two things lead to genes' being "successful": (1) the organism is able to successfully mate, passing its genes on, and (2) its offspring are able to survive in order to do the same.

Darwin recognized this implication of his theory of natural selection and came up with the idea of sexual selection. Because an organism must reproduce to pass its genes on, qualities that will attract a mate should eventually become encoded in the genome. If a square jaw and outsized biceps are attractive features for a man to have (in the eyes of potential mates), men with those features will reproduce more successfully than their narrow-jawed, scrawny-armed competitors. The square-jaw, large-bicep genes will then become more plentiful. Offspring also need to be protected from the elements, from predators, from disease, and to be given food and other resources so that they can reproduce. Thus, a gene that promotes nurturing behavior postcopulation could also spread throughout the population, to the extent that the offspring of people with the nurturing gene fare better, as a group, in the competition for resources and mates.

Might music play a role in sexual selection? Darwin thought so. In *The Descent of Man* he wrote, "I conclude that musical notes and rhythm were first acquired by the male or female progenitors of mankind for the sake of charming the opposite sex. Thus musical tones became firmly associated with some of the strongest passions an animal is capable of feeling, and are consequently used instinctively. . . ." In seeking mates, our innate drive is to find—either consciously or unconsciously—someone who is biologically and sexually fit, someone who will provide us with children who are likely to be healthy and able to attract mates of their own. Music may indicate biological and sexual fitness, serving to attract mates.

Darwin believed that music preceded speech as a means of courtship, equating music with the peacock's tail. In his theory of sexual selection, Darwin posited the emergence of features that served no direct survival purpose other than to make oneself (and hence one's genes) attractive. The cognitive psychologist Geoffrey Miller has connected this notion with the role that music plays in contemporary society. Jimi Hendrix had "sexual liaisons with hundreds of groupies, maintained parallel long-term relationships with at least two women, and fathered at least three children in the United States, Germany, and Sweden. Under ancestral conditions before birth control, he would have fathered many more," Miller writes. Robert Plant, the lead singer of Led Zeppelin, recalls his experience with their big concert tours in the seventies:

"I was on my way to love. Always. Whatever road I took, the car was heading for one of the greatest sexual encounters I've ever had."

The number of sexual partners for rock stars can be hundreds of times what a normal male has, and for the top rock stars, such as Mick Jagger, physical appearance doesn't seem to be an issue.

During sexual courtship, animals often advertise the quality of their genes, bodies, and minds, in order to attract the best possible mate. Many human-specific behaviors (such as conversation, music production, artistic ability, and humor) may have evolved principally to advertise intelligence during courtship. Miller suggests that under the conditions that would have existed throughout most of our evolutionary history in which music and dance were completely intertwined, musicianship/danceship would have been a sign of sexual fitness on two fronts. First, anyone who could sing and dance was advertising to potential mates his stamina and overall good health, physical and mental. Second, anyone who had become expert or accomplished in music and dance was advertising that he had enough food and sturdy enough shelter that he could afford to waste valuable time on developing a purely unnecessary skill. This is the argument of the peacock's beautiful tail: The size of the peacock's tail correlates with the bird's age, health, and overall fitness. The colorful tail signals that the healthy peacock has metabolism to waste, he is so fit, so together, so wealthy (in terms of resources) that he has extra resources to put into something that is purely for display and aesthetic purposes.

In contemporary society, we see this with rich people who build elaborate houses or drive hundred-thousand-dollar cars. The sexual selection message is clear: Choose me. I have so much food and so many resources that I can afford to squander them on these luxury items. It is no accident that many men living at or near the poverty line in the U.S. buy old Cadillacs and Lincolns—impractical, high-status vehicles that unconsciously signal their owner's sexual fitness. This can also be seen as the origin of bling, the tendency for men to wear gaudy jewelry. That the yearning for and purchasing of cars and jewelry peaks in men during adolescence, when they are most sexually potent, serves the theory. Music making, because it involves an array of physical and mental skills, would be an overt display of health, and to the extent that someone had time to develop his musicianship, the argument goes, it would indicate resource wealth.

In contemporary society, interest in music also peaks during adolescence, further bolstering the sexual-selection aspects of music. Far more nineteen-year-olds are starting bands and trying to get their hands on new music than are forty-year-olds, even though the forty-year-olds have had more time to develop their musicianship and preferences. "Music evolved and continues to function as a courtship display, mostly broadcast by young males to attract females," Miller argues.

Music as a sexual fitness display is not so farfetched an idea when we realize the form that hunting took in some hunter-gatherer societies. Some protohumans would rely on persistence hunting—hurling spears, rocks, and other projectiles at their prey, then chasing the prey for hours until the animal dropped from injury and exhaustion. If dancing in past hunter-gatherer societies was anything like what we observe in contemporary ones, it typically extended for many hours, requiring great aerobic effort. As a display of a male's fitness to take part in or lead a hunt, such tribal dancing would be an excellent indicator. Most tribal dancing includes repeated high-stepping, stomping, and jumping using the largest, most energy-hungry muscles of the body. Many mental illnesses are now known to undermine the ability to dance or to perform rhythmically—schizophrenia and Parkinson's, to name just two—and so the sort of rhythmic dancing and music making that have characterized most music across the ages serves as a warranty of physical and mental fitness, perhaps even a warranty of reliability and conscientiousness (because, as we saw in Chapter 7, expertise requires a particular kind of mental focus).

Another possibility is that evolution selected creativity in general as a marker of sexual fitness. Improvisation and novelty in a combined music/dance performance would indicate the cognitive flexibility of the dancer, signaling his potential for cunning and strategizing while on the hunt. The material wealth of a male suitor has long been considered among the most compelling attractors to females, who assume that it will increase the likelihood of their offspring having ample food, shelter, and protection. (Protection accrues to the wealthy because they can marshal support of other community members in exchange for food or symbolic tokens of wealth such as jewelry or cash.) If wealth is the name of the dating game, then music would seem relatively unimportant. But Miller and his colleague Martie Haselton at UCLA have shown that creativity trumps wealth, at least in human females. Their hypothesis is that while wealth may predict who will make a good dad (for child rearing), creativity may better predict who will furnish the best genes (for child fathering).

In a clever study, women at various stages of their normal menstrual cycle—some during their peak of fertility, others at their minimum of fertility and others in between—were asked to rate the attractiveness of potential mates based on written vignettes describing fictional males. A typical vignette described a man who was an artist, and who displayed great creative intelligence in his work, but who was poor due to bad luck. An alternate vignette described a man who had average creative intelligence, but happened to be wealthy due to good luck. All the vignettes were designed to make clear that each man's creativity was a function of his traits and attributes (and thus, endogenous, genetic, and heritable) while each man's financial state was largely accidental (and thus exogenous and not heritable).

The results showed that when they were at their peak fertility, women preferred the creative but poor artist to the not creative but rich man as a short-term mate, or for a brief sexual encounter. At other times during their cycle, women did not show such preferences. It is important to bear in mind that preferences are to a large degree hardwired and not easily overpowered by conscious cognitions; the fact that women today can avoid pregnancy through almost foolproof birth control is a concept so new in our species as to have no influence on any innate preferences. The men (and women) who might make the best caregivers are not necessarily those who can contribute the best genes to potential offspring. People don't always marry those to whom they are the most sexually attracted, and 50 percent of people of both sexes report to having extramarital affairs. Far more women want to sleep with rock stars and athletes than to marry them. In short, the best fathers (in the biological sense) don't always make the best dads (for child rearing). This may account for why, according to a recent European study, 10 percent of mothers reported that their children were being raised by men who falsely believed the children were their own. Although today reproduction may not be the motive, it is difficult to separate out innate, evolutionarily derived preferences for mating partners from our societally and culturally induced tastes in sexual partners.

For musicologist David Huron of Ohio State, the key question for the evolutionary basis is what advantage might be conferred on individuals who exhibit musical behaviors, versus those who do not. If music is a nonadaptive pleasure-seeking behavior—the auditory cheesecake argument—we would expect it not to last very long in evolutionary time. Huron writes, "Heroin users tend to neglect their health and are known to have high mortality rates. Furthermore, heroin users make poor parents; they tend to neglect their offspring." Neglecting one's health and the health of one's children is a surefire way to reduce the probability of one's genes being passed on to future generations. First, if music was nonadaptive, then music lovers should be at some evolutionary or survival disadvantage. Second, music shouldn't have been around very long. Any activity that has low adaptive value is unlikely to be practiced for very long in the species's history, or to occupy a significant portion of an individual's time and energy.

All the available evidence is that music can't be merely auditory cheesecake; it has been around a very long time in our species. Musical instruments are among the oldest human-made artifacts we have found. The Slovenian bone flute, dated at fifty thousand years ago, which was made from the femur of a now-extinct European bear, is a prime example. Music predates agriculture in the history of our species. We can say, conservatively, that there is no tangible evidence that language preceded music. In fact, the physical evidence suggests the contrary. Music is no doubt older than the fifty-thousand-year-old bone flute, because flutes were unlikely the first instruments. Various percussion instruments, including drums, shakers, and rattles were likely to have been in use for thousands of years before flutes—we see this in contemporary hunter-gatherer societies, and from the record of European invaders reporting on what they found in native American cultures. The archaeological record shows an uninterrupted record of music making everywhere we find humans, and in every era. And, of course, singing most probably predated flutes as well.

To restate the summary principle of evolutionary biology, "Genetic mutations that enhance one's likelihood to live long enough to reproduce become adaptations." The best estimates are that it takes a minimum of fifty thousand years for an adaptation to show up in the human genome. This is called evolutionary lag—the time lag between when an adaptation first appears in a small proportion of individuals and when it becomes widely distributed in the population. When behavioral geneticists and evolutionary psychologists look for an evolutionary explanation for our behaviors or appearance, they consider what evolutionary problem was being addressed by the adaptation in question. But due to evolutionary lag, the adaptation in question would have been a response to conditions as they were at least fifty thousand years ago, not as they are today. Our hunter-gatherer ancestors had a very different lifestyle than anyone who is reading this book, with different priorities and pressures. Many of the problems we face today—cancers, heart disease, maybe even the high divorce rate—have come to torment us because our bodies and our brains were designed to handle life the way it was for us fifty thousand years ago. Fifty thousand years from now in the year 52,006 (give or take a few millennia), our species may finally have evolved to handle life the way it is now, with overcrowded cities, air and water pollution, video games, polyester, glazed doughnuts, and a gross imbalance in the distribution of resources worldwide. We may evolve mental mechanisms that allow us to live in close quarters without feeling a loss of privacy, and physiological mechanisms to process carbon monoxide, radioactive waste, and refined sugar, and we may learn to use resources that today are unusable.

When we ask about the evolutionary basis for music, it does no good to think about Britney or Bach. We have to think what music was like around fifty thousand years ago. The instruments recovered from archeological sites can help us understand what our ancestors used to make music, and what kinds of melodies they listened to. Cave paintings, paintings on stoneware, and other pictorial artifacts can tell us something about the role that music played in daily life. We can also study contemporary societies that have been cut off from civilization as we know it, groups of people who are living hunter-gatherer lifestyles that have remained unchanged for thousands of years. One striking find is that in every society of which we're aware, music and dance are inseparable.

The arguments against music as an adaptation consider music only as disembodied sound, and moreover, as performed by an expert class for an audience. But it is only in the last five hundred years that music has become a spectator activity—the thought of a musical concert in which a class of "experts" performed for an appreciative audience was virtually unknown throughout our history as a species. And it has only been in the last hundred years or so that the ties between musical sound and human movement have been minimized. The embodied nature of music, the indivisibility of movement and sound, the anthropologist John Blacking writes, characterizes music across cultures and across times. Most of us would be shocked if audience members at a symphonic concert got out of their chairs and clapped their hands, whooped, hollered, and danced as is de rigueur at a James Brown concert. But the reaction to James Brown is certainly closer to our true nature. The polite listening response, in which music has become an entirely cerebral experience (even music's emotions are meant, in the classical tradition, to be felt internally and not to cause a physical outburst) is counter to our evolutionary history. Children often show the reaction that is true to our nature: Even at classical music concerts they sway and shout and generally participate when they feel like it. We have to train them to behave "civilized."

When a behavior or trait is widely distributed across members of a species, we take it to be encoded in the genome (regardless of whether it was an adaptation or a spandrel). Blacking argues that the universal distribution of music-making ability in African societies suggests that "musical ability [is] a general characteristic of the human species, rather than a rare talent." More important, Cross writes that "musical ability cannot be defined solely in terms of productive competence"; virtually every member of our own society is capable of listening to and hence of understanding music.

Apart from these facts about music's ubiquity, history, and anatomy, it is important to understand how and why music was selected. Darwin proposed the sexual-selection hypothesis, which has been advanced more recently by Miller and others. Additional possibilities have been argued as well. One is social bonding and cohesion. Collective music making may encourage social cohesions—humans are social animals, and music may have historically served to promote feelings of group togetherness and synchrony, and may have been an exercise for other social acts such as turn-taking behaviors. Singing around the ancient campfire might have been a way to stay awake, to ward off predators, and to develop social coordination and social cooperation within the group. Humans need social linkages to make society work, and music is one of them.

An intriguing line of evidence for the social-bonding basis of music comes from my work with Ursula Bellugi on individuals with mental disorders such as Williams syndrome (WS) and autism spectrum disorders (ASD). As we saw in Chapter 6, WS is genetic in origin, and causes abnormal neuronal and cognitive development, resulting in intellectual impairment. People with WS, in spite of their overall mental impairment, are particularly good at music, and they're particularly social.

A contrast is people with ASD, many of whom also suffer from intellectual impairment. It remains a controversial issue whether ASD has a genetic basis or not. A marker of ASD is the inability to empathize with others, to understand emotions or emotional communication, particularly emotions in others. People with ASD can certainly become angry and upset, they are not robots. But their ability to "read" the emotions of others is significantly impaired, and this typically extends to their utter inability to appreciate the aesthetic qualities of art and music. Although some people with ASD play music, and some of them have reached a high level of technical proficiency, they do not report being emotionally moved by music. Rather, the preliminary and largely anecdotal evidence is that they are attracted to the *structure* of music. Temple Grandin, a professor who is autistic, has written that she finds music "pretty" but that in general, she just "doesn't get it" or understand why people react to it the way that they do.

With WS and ASD, we have two complementary syndromes. On the one hand we have a population who are highly social, gregarious, and highly musical. On the other, we have a population who are highly antisocial and not very musical. The putative link between music and social bonding is strengthened by complementary cases such as these, what neuroscientists call a double dissociation. The argument is that there may be a cluster of genes that influences both outgoingness and musicality. If this were true, we would expect to find that deviations in one ability co-occur with deviations in the other, as we do in WS and ASD.

The brains of people with WS and ASD also, as we might expect, reveal complementary impairments. Allan Reiss has shown that the neocerebellum, the newest part of the cerebellum, is larger than normal in WS, and smaller than normal in ASD. Because we already know the important role played by the cerebellum in music cognition, this is not surprising. Some as yet unidentified genetic abnormality appears to cause, either directly or indirectly, the neural dismorphology in WS, and we presume in ASD as well. This, in turn, leads to abnormal development of musical behaviors that in one case are enhanced and the other are diminished.

Because of the complex and interactive nature of genes, it is certain that there are other genetic correlates to sociability and musicality that go beyond the cerebellum. The geneticist Julie Korenberg has speculated that there exists a cluster of genes that are related to outgoingness versus inhibitedness, and that people with WS lack some of the normal inhibition genes that the rest of us have, causing their musical behaviors to be more uninhibited; for over a decade anecdotal reports, on CBS News's *60 Minutes*, in a movie narrated by Oliver Sacks on Williams, and in a host of newspaper articles, have claimed that people with WS are more fully engaged with—immersed in—music than most people. My own laboratory has provided neural evidence for this point. We scanned the brains of individuals with WS while they listened to music, and found they were using a vastly larger set of neural structures than everyone else does. Activation in their amygdala and cerebellum—the emotional centers of the brain—was significantly stronger than in "normals." Everywhere we looked, we found stronger neural activation, and more widespread neural activation. Their brains were humming.

A third argument in favor of music's primacy in human (and proto-human) evolution is that music evolved because it promoted cognitive development. Music may be *the* activity that prepared our pre-human ancestors for speech communication and for the very cognitive, representational flexibility necessary to become humans. Singing and instrumental activities might have helped our species to refine motor skills, paving the way for the development of the exquisitely fine muscle control required for vocal or signed speech. Because music is a complex activity, Trehub suggests that it may help prepare the developing infant for its mental life ahead. It shares many of the features of speech and it may form a way of "practicing" speech perception in a separate context. No human has ever learned language by memorization. Babies don't simply memorize every word and sentence they've ever heard; rather, they learn rules and apply them in perceiving and generating new speech. One piece of evidence for this is empirical; the other is logical. The empirical evidence comes from what linguists call overextension: Children just learning the rules of language often apply them logically, but incorrectly. We see this most clearly in the case of irregular verb conjugations and irregular plurals in English. The developing brain is primed to make new neural connections and to prune away old ones that are not useful or accurate, and its mission is to instantiate rules insofar as possible. This is why we hear young children say, "He goed to the store," instead of "He went to the store." They are applying a logical rule: Most English verbs in past tense take an *-ed* ending: *play/played, talk/talked, touch/touched.* Reasonable application of the rule leads to overextensions such as *buyed, swimmed,* and *eated.* In fact, intelligent children are more likely to make these mistakes and to make them sooner during the course of their development, because they have a more sophisticated rule-generating system. Because many, many children make these speech errors and few adults do, this is evidence that children are not simply mimicking what they hear, but rather, their brains are developing theories and rules about speech that they then apply.

The second piece of evidence that children don't simply memorize language is logical: All of us speak sentences that we've never heard before. We can form an infinite number of sentences to express thoughts and ideas that we have neither expressed before nor heard expressed before—that is, language is generative. Children must learn the grammatical rules for generating unique sentences to become competent speakers of their language. A trivial example of how the number of sentences in human language is infinite is that for any sentence you give me, I can always add "I don't believe" to the beginning of it, and make a new sentence. "I like beer" becomes "I don't believe I like beer." "Mary says she likes beer" becomes "I don't believe Mary says she likes beer." Even "I don't believe Mary says she likes beer" becomes "I don't believe I don't believe Mary says she likes beer." Although a sentence like this is awkward, it doesn't alter the fact that it expresses a new idea. For language to be generative, children must not be learning by rote. Music is also generative. For every musical phrase I hear, I can always add a note to the beginning, end, or middle to generate a new musical phrase.

Cosmides and Tooby argue that music's function in the developing child is to help prepare its mind for a number of complex cognitive and social activities, exercising the brain so that it will be ready for the demands placed on it by language and social interaction. The fact that music lacks specific referents makes it a safe symbol system for expressing mood and feelings in a nonconfrontational manner. Music processing helps infants to prepare for language; it may pave the way to linguistic prosody, even before the child's developing brain is ready to process phonetics. Music for the developing brain is a form of play, an exercise that invokes higher-level integrative processes that nurture exploratory competence, preparing the child to eventually explore generative language development through babbling, and ultimately more complex linguistic and paralinguistic productions.

Mother-infant interactions involving music almost always entail both singing and rhythmic movement, such as rocking or caressing. This appears to be culturally universal. During the first six months or so of life, as I showed in Chapter 7, the infant brain is unable to clearly distinguish the source of sensory inputs; vision, hearing, and touch meld into a unitary perceptual representation. The regions of the brain that will eventually become the auditory cortex, the sensory cortex, and the visual cortex are functionally undifferentiated, and inputs from the various sensory receptors may connect to many different parts of the brain, pending pruning that will occur later in life. As Simon Baron-Cohen has described it, with all this sensory cross talk, the infant lives in a state of complete psychedelic splendor (without the aid of drugs).

Cross explicitly acknowledges that what music has become, today, with the influence of time and culture, is not necessarily what it was fifty thousand years ago, nor should we expect it to be. But considering ancient music's character does account for why so many of us are literally

moved by rhythm; by almost all accounts the music of our distant ancestors was heavily rhythmic. Rhythm stirs our bodies. Tonality and melody
stir our brains. The coming together of rhythm and melody bridges our
cerebellum (the motor control, primitive little brain) and our cerebral
cortex (the most evolved, most human part of our brain). This is how
Ravel's *Bolero*, Charlie Parker's "Koko," or the Rolling Stones' "Honky
Tonk Women" inspire us and move us, both metaphorically and physically, exquisite unions of time and melodic space. It is why rock, metal,
and hip-hop music are the most popular musical genres in the world, and
have been for the past twenty years. Mitch Miller, the head talent scout
for Columbia Records, famously said in the early sixties that rock-and-
roll music was a fad that would soon die. Now, in 2006, there is no sign
of it slowing down. Classical music as most of us think of it—say, from
1575 to 1950, from Monteverdi to Bach to Stravinsky, Rachmaninoff, and
so on—is no longer being written. Contemporary composers in music
conservatories are not creating this sort of music as a rule, but rather,
they are writing what many refer to as twentieth-century (now twenty-
first-century) art music. And so we have Philip Glass and John Cage and
more recent, lesser-known composers whose music is rarely performed
by our symphony orchestras. When Copeland and Bernstein were composing, orchestras played their works and the public enjoyed them. This
seems to be less and less the case in the past forty years. Contemporary
"classical" music is practiced mostly in universities; it is listened to by almost no one; it deconstructs harmony, melody, and rhythm, rendering
them all but unrecognizable; it is a purely intellectual exercise, and save
for the rare avant-garde ballet company, no one dances to it either.

A fourth argument for music as an adaptation comes from other
species. If we can show that other species use music for similar purposes, this presents a strong evolutionary argument. It is especially
important, however, not to anthropomorphize animal behaviors, interpreting them only from our own cultural perspective. What sounds to us
like music or a song may be serving, in animals, a very different function
for them than it does for us. When we see a dog rolling around in fresh
summer grass, with that very doglike grin on his face, we think, "Spike

must be really happy." We are interpreting the rolling-on-the-grass behavior in terms of what we know about our own species, without stopping to consider that it might mean something different to Spike and to
his species. Human children roll around in the grass, do somersaults and
cartwheels, when they are happy. Male dogs roll around in the grass
when they find a particularly pungent smell there, preferably from a recently dead animal, and they cover their fur with it to make other dogs
think that they are skilled hunters. Similarly, birdsong that sounds joyful
to us is not necessarily intended that way by the bird-singer, or interpreted that way by the bird-listener.

Yet of all the calls of other species, birdsong holds a special position
of awe and intrigue. Who among us hasn't sat and listened to a songbird
on a spring morning and found the beauty, the melody, the structure of
it enticing? Aristotle and Mozart were among those who did; they considered the songs of a bird to be every bit as musical as the compositions
of humans. But why do we write and perform music? Are our motiva-
tions any different from those of the animals?

Birds, whales, gibbons, frogs, and other species use vocalizations for
a variety of purposes. Chimpanzees and prairie dogs have alert calls to
caution their fellows about an approaching predator, and the calls are
specific to the predator. Chimps use one vocalization to signal an approaching eagle (alerting their primate pals to hide underneath something) and another to broadcast the incursion of a snake (alerting their
friends to climb a tree). Male birds use their vocalizations to establish
territory; robins and crows reserve a particular call to warn of predators
such as dogs and cats.

Other animal vocalizations are more clearly related to courtship. In
songbirds, it is generally the male of the species that sings, and for some
species, the larger the repertoire, the more likely he is to attract a mate.
Yes, for a female songbird, size matters; it indicates male-bird intellect
and, by extension, a source of potentially good bird genes. This was
shown in a study that played different songs over loudspeakers to female
birds. The birds ovulated more quickly in the presence of a large birdsong
repertoire than in the presence of a small one. Some male songbirds will

sing their courtship song until they drop dead from exhaustion. Linguists point to the generative nature of human music, the ability we have to create new songs out of components, in an almost limitless fashion. This is not a uniquely human trait either. Several bird species generate their songs from a repertoire of basic sounds, creating new melodies and variations on them, and the male who sings the most elaborate songs is typically the one who is most successful at mating. Music's function in sexual selection thus has an analogue in other species.

Music's evolutionary origin is established because it is present across all humans (meeting the biologists' criterion of being widespread in a species); it has been around a long time (refuting the notion that it is merely audio cheesecake); it involves specialized brain structures, including dedicated memory systems that can remain functional when other memory systems fail (when a physical brain system develops across all humans, we assume that it has an evolutionary basis); and it is analogous to music making in other species. Rhythmic sequences optimally excite recurrent neural networks in mammalian brains, including feedback loops among the motor cortex, the cerebellum, and the frontal regions. Tonal systems, pitch transitions, and chords scaffold on certain properties of the auditory system that were themselves products of the physical world, of the inherent nature of vibrating objects. Our auditory system develops in ways that play on the relation between scales and the overtone series. Musical novelty attracts attention and overcomes boredom, increasing memorability.

Darwin's theory of natural selection was revolutionized by the discovery of the gene, specifically Watson and Crick's discovery of the structure of DNA. Perhaps we are witnessing another revolution in the aspect of evolution that depends on social behavior, on culture.

Undoubtedly one of the most cited discoveries in neuroscience in the past twenty years was of mirror neurons in the primate brain. Giacomo Rizzolatti, Leonardo Fogassi, and Vittorio Gallese were studying the the brain mechanisms responsible for movements such as reaching and grasping in monkeys. They read the output from a single neuron in the monkey's brain as it reached for pieces of food. At one point, Fogassi reached for a banana, and the monkey's neuron—one that had already been associated with movement—started to fire. "How could this happen, when the monkey did not move?" Rizzolatti recalls thinking. "At first we thought it to be a flaw in our measuring or maybe equipment failure, but everything checked out OK and the reaction was repeated as we repeated the movement." A decade of work since then has established that primates, some birds, and humans have mirror neurons, neurons that fire both when performing an action and when observing someone else performing that action.

The purpose of mirror neurons is presumably to train and prepare the organism to make movements it has not made before. We've found mirror neurons in Broca's area, a part of the brain intimately involved in speaking, and in learning to speak. Mirror neurons may explain an old mystery of how it is that infants learn to imitate the faces that parents make at them. It may also explain why musical rhythm moves us, both emotionally and physically. We don't yet have solid evidence, but some neuroscientists speculate that our mirror neurons may be firing when we see or hear musicians perform, as our brain tries to figure out how those sounds are being created, in preparation for being able to mirror or echo them back as part of a signaling system. Many musicians can play back a musical part on their instruments after they've heard it only once. Mirror neurons are likely involved in this ability.

Genes are what pass protein recipes between individuals and across generations. Maybe mirror neurons, now in concert with sheet music, CDs, and iPods, will turn out to be the fundamental messengers of music across individuals and generations, enabling that special kind of evolution—cultural evolution—through which develop our beliefs, obsessions, and all of art.

For many solitary species, the ability to ritualize certain aspects of fitness in a courtship display makes sense, because a potential mate pair may only see each other for a few minutes. But in highly social societies like ours, why would you need to demonstrate fitness through such a highly stylized and symbolic means as dancing and singing? Humans live in social groups and have ample opportunities to observe one another in a variety of situations and over long periods of time. Why would music be needed to show fitness? Primates are highly social, living in groups, forming complex long-term relationships that involve social strategies. Hominid courtship was probably a long-term affair. Music, particularly memorable music, would insinuate itself into the mind of a potential mate, leading her to think about her suitor even when he was out on a long hunt, and predisposing her toward him when he returned. The multiple reinforcing cues of a good song—rhythm, melody, contour—cause music to stick in our heads. That is the reason that many ancient myths, epics, and even the Old Testament were set to music in preparation for being passed down by oral tradition across the generations. As a tool for activation of specific thoughts, music is not as good as language. As a tool for arousing feelings and emotions, music is better than language. The combination of the two—as best exemplified in a love song—is the best courtship display of all.

APPENDIX A

This Is Your Brain on Music

Music processing is distributed throughout the brain. The figures on the following two pages show the brain's major computational centers for music. The first illustration is a view of the brain from the side. The front of the brain is to the left. The second illustration shows the inside of the brain from the same point of view as the first illustration. These figures are based on illustrations by Mark Tramo published in *Science* in 2001, but are redrawn and include newer information.

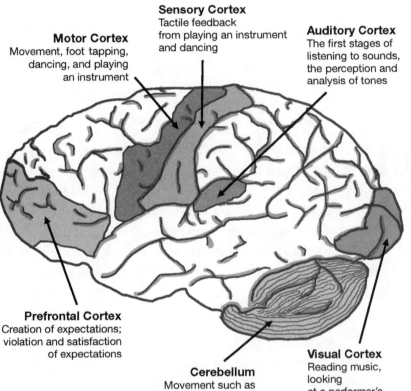

Motor Cortex
Movement, foot tapping,
dancing, and playing
an instrument

Sensory Cortex
Tactile feedback
from playing an instrument
and dancing

Auditory Cortex
The first stages of
listening to sounds,
the perception and
analysis of tones

Prefrontal Cortex
Creation of expectations;
violation and satisfaction
of expectations

Cerebellum
Movement such as
foot tapping, dancing,
and playing an instrument.
Also involved in emotional
reactions to music

Visual Cortex
Reading music,
looking
at a performer's
movements
(including one's own)

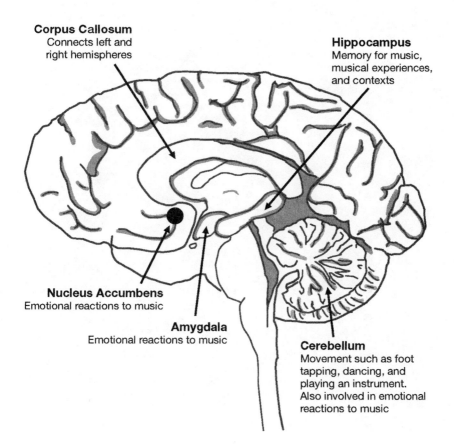

Corpus Callosum
Connects left and
right hemispheres

Hippocampus
Memory for music,
musical experiences,
and contexts

Nucleus Accumbens
Emotional reactions to music

Amygdala
Emotional reactions to music

Cerebellum
Movement such as foot
tapping, dancing, and
playing an instrument.
Also involved in emotional
reactions to music

APPENDIX B

Chords and Harmony

Within the key of C, the only legal chords are chords built off of the notes of the C major scale. This causes some chords to be major and some minor, because of the unequal spacing of tones in the scale. To build the standard three-note chord—a triadic chord—we start on any of the tones of the C major scale, skip one, and then use the next, then skip one again and use the next one after that. The first chord in C major, then, comes from the notes C-E-G, and because the first interval formed, between C and E, is a major third, we call this chord a major chord, and in particular, a C major chord. The next one we build in a similar fashion is composed of D-F-A. Because the interval between D and F is a minor third, this chord is called a D minor chord. Remember, major chords and minor chords have a very different sound. Even though most nonmusicians can't name a chord on hearing it, or label it as major or minor, if they hear a major and minor chord back to back they can tell the difference. And their brains can certainly tell the difference—a number of studies have shown that nonmusicians produce different physiological responses to major versus minor chords, and major versus minor keys.

In the major scale, considering the triadic chords constructed in the standard way I've just described, three are major (on the first, fourth, and fifth scale degrees), three are minor (on the second, third, and sixth degrees) and one is called a diminished chord (on the seventh scale degree) and is made up of two intervals of a minor third. The reason we say that we're in the key of C major, even though there are three minor chords in the key, is because the root chord—the chord that the music points to, the one that feels like "home"—is C major.

Generally, composers use chords to set a mood. The use of chords and the way they are strung together is called harmony. Another, perhaps better-known use of the word *harmony* is to indicate when two or more singers or instrumentalists are playing together and they're not playing the same notes, but conceptually this is the same idea. Some chord sequences are used more than others, and can become typical of a particular genre. For example, the blues is defined by a particular chord sequence: a major chord on the first scale degree (written I major) followed by a major chord on the fourth scale degree (written IV major) followed by I major again, then V major, optionally to IV major, then back to I major. This is the standard blues progression, found in songs such as "Crossroads" (Robert Johnson, later covered by Cream), "Sweet Sixteen" by B. B. King, and "I Hear You Knockin'" (as recorded by Smiley Lewis, Big Joe Turner, Screamin' Jay Hawkins, and Dave Edmunds). The blues progression—either verbatim or with some variations—is the basis for rock and roll music, and is found in thousands of songs including "Tutti Frutti" by Little Richard, "Rock and Roll Music" by Chuck Berry, "Kansas City," by Wilbert Marrison, "Rock and Roll" by Led Zeppelin, "Jet Airliner" by the Steve Miller Band (which is surprisingly similar to "Crossroads"), and "Get Back" by the Beatles. Jazz artists such as Miles Davis and progressive rock artists like Steely Dan have written dozens of songs that are inspired by this progression, with their own creative ways of substituting more exotic chords for the standard three; but they are still blues progressions, even when dressed up in fancier chords.

Bebop music leaned heavily on a particular progression originally written by George Gershwin for the song "I've Got Rhythm." In the key of C, the basic chords would be:

C major–A minor–D minor–G7–C major–A minor–D minor–G7

C major–C7–F major–F minor–C major–G7–C major

C major–A minor–D minor–G7–C major–A minor–D minor–G7

C major–C7–F–F minor–C major–G7–C major

The 7 next to a note name indicates a tetrad—a four-note chord—that is simply a major chord with a fourth note added on top; the top note is a minor third above the third note of the chord. The chord G7 is called either "G seven" or "G dominant seven." Once we start using tetrads instead of triads for chords, a great deal of rich tonal variation is possible. Rock and blues tend to use only the dominant seven, but there are two other types of "seven" chords in common use, each conveying a different emotional flavor. "Tin Man" and "Sister Golden Hair" by the group America use the major seven chord to give them their characteristic sound (a major triad with a major third on top, rather than the minor third of the chord we're calling the dominant seven); "The Thrill Is Gone" by B. B. King uses minor seven chords throughout (a minor triad with a minor third on top).

The dominant seven chord occurs naturally—that is, diatonically—when it starts on the fifth degree of the major scale. In the key of C, then, G7 can be constructed by playing all white notes. The dominant seven contains that formerly banned interval, the tritone, and it is the only chord in a key that does. The tritone is harmonically the most unstable interval we have in Western music, and so it carries with it a very strong perceptual urge to resolve. Because the dominant seven chord also contains the most unstable scale tone—the seventh degree (B in the key of C)—the chord "wants to" resolve back to C, the root. It is for this reason that the dominant seven chord built on the fifth degree of a major scale—the V7 chord, or G7 in the key of C—is the most typical, standard, and clichéd chord right before a composition ends on its root. In other words, the combination of G7 to C major (or their equivalents in other keys) gives us the single most unstable chord followed by the single most stable chord; it gives us the maximum feeling of tension and resolution that we can have. At the end of some of Beethoven's symphonies, when the ending seems to go on and on and on, what the maestro is doing is giving us that two-chord progression over and over and over again until the piece finally resolves on the root.

BIBLIOGRAPHIC NOTES

The following are some of the many articles and books that I have consulted. The list is by no means complete, but represents additional sources that are most relevant to the points made in this book. This book was written for the non-specialist and not for my colleagues, and so I have tried to simplify topics without oversimplifying them. A more complete and detailed account of the brain and music can be found in these readings, and in the readings cited in them. Some of the works cited below are written for the specialist researcher. I have used an asterisk (*) to indicate the more technical readings. Most of the marked entries are primary sources, and a few are graduate-level textbooks.

Introduction

Churchland, P. M. 1986. *Matter and Consciousness*. Cambridge: MIT Press.
> In the passage on mankind's curiosity having solved many of the greatest scientific mysteries, I have borrowed liberally from the introduction to this excellent and inspiring work on the philosophy of mind.

*Cosmides, L., and J. Tooby. 1989. Evolutionary psychology and the generation of culture, Part I. Case study: A computational theory of social exchange. *Ethology and Sociobiology* 10: 51–97.
> An excellent introduction to the field of evolutionary psychology by two leading scholars.

*Deaner, R. O., and C. L. Nunn. 1999. How quickly do brains catch up with bodies? A comparative method for detecting evolutionary lag. *Proceedings of Biological Sciences* 266 (1420):687–694.
> A recent scholarly article on the topic of evolutionary lag, the notion that

our bodies and minds are at present equipped to deal with the world and living conditions as they were fifty thousand years ago, due to the amount of time it takes for adaptations to become encoded in the human genome.

Levitin, D. J. 2001. Paul Simon: The Grammy Interview. *Grammy* September, 42–46.
> Source of the Paul Simon quote about listening for sound.

*Miller, G. F. 2000. Evolution of human music through sexual selection. In *The Origins of Music*, edited by N. L. Wallin, B. Merker, and S. Brown. Cambridge: MIT Press.
> Written by another leader in the field of evolutionary psychology, this article discusses many of the ideas discussed in Chapter 9, which are mentioned only briefly in Chapter 1.

Pareles, J., and P. Romanowski, eds. 1983. *The Rolling Stone Encyclopedia of Rock & Roll*. New York: Summit Books.
> Adam and the Ants get eight column inches plus a photo in this edition, U2—already well known with three albums and the hit "New Year's Day"—get only four inches, and no photo.

*Pribram, K. H. 1980. Mind, brain, and consciousness: the organization of competence and conduct. In *The Psychobiology of Consciousness*, edited by J. M. D. Davidson, R.J. New York: Plenum.

*———. 1982. Brain mechanism in music: prolegomena for a theory of the meaning of meaning. In *Music, Mind, and Brain*, edited by M. Clynes. New York: Plenum.
> Pribram taught his course from a collection of articles and notes that he had compiled. These were two of the papers that we read.

Sapolsky, R. M. 1998. *Why Zebras Don't Get Ulcers*, 3rd ed. New York: Henry Holt and Company.
> An excellent book and a fun read on the science of stress, and the reasons that modern humans suffer from stress; the idea of "evolutionary lag" that I introduce more fully in Chapter 9 is dealt with very well in this book.

*Shepard, R. N. 1987. Toward a Universal Law of Generalization for psychological science. *Science* 237 (4820):1317–1323.

*———. 1992. The perceptual organization of colors: an adaptation to regularities of the terrestrial world? In *The Adapted Mind: Evolutionary Psychology and the Generation of Culture*, edited by J. H. Barkow, L. Cosmides, and J. Tooby. New York: Oxford University Press.

*———. 1995. Mental universals: Toward a twenty-first century science of mind. In *The Science of the Mind: 2001 and Beyond*, edited by R. L. Solso and D. W. Massaro. New York: Oxford University Press.
> Three papers by Shepard in which he discusses the evolution of mind.

Tooby, J., and L. Cosmides. 2002. Toward mapping the evolved functional organization of mind and brain. In *Foundations of Cognitive Psychology*, edited by D. J. Levitin. Cambridge: MIT Press.
> Another paper by these two leaders in evolutionary psychology, perhaps the more general of the two papers I've listed here.

Chapter 1

*Balzano, G. J. 1986. What are musical pitch and timbre? *Music Perception* 3 (3):297–314.
> A scientific article on the issues involved in pitch and timbre research.

Berkeley, G. 1734/2004. *A Treatise Concerning the Principles of Human Knowledge*. Whitefish, Mont.: Kessinger Publishing Company.
> The famous question—if a tree falls in the forest and no one is there to hear it, does it make a sound?—was first posed by the theologian and philosopher George Berkeley, bishop of Cloyne, in this work.

*Bharucha, J. J. 2002. Neural nets, temporal composites, and tonality. In *Foundations of Cognitive Psychology: Core Readings*, edited by D. J. Levitin. Cambridge: MIT Press.
> Neural networks for chord recognition.

*Boulanger, R. 2000. *The C-Sound Book: Perspectives in Software Synthesis, Sound Design, Signal Processing, and Programming*. Cambridge: MIT Press.
> An introduction to the most widely used software sound synthesis program/system. The best book I know of for people who want to learn to program computers to make music and create timbres of their own choosing.

Burns, E. M. 1999. Intervals, scales, and tuning. In *Psychology of Music*, edited by D. Deutsch. San Diego: Academic Press.
> On the origin of scales, relationships among tones, nature of intervals and scales.

*Chowning, J. 1973. The synthesis of complex audio spectra by means of frequency modulation. *Journal of the Audio Engineering Society* 21:526–534.
> FM synthesis, as eventually manifested in the Yamaha DX synthesizers, was first described in this professional journal.

Clayson, A. 2002. *Edgard Varèse*. London: Sanctuary Publishing, Ltd.
> Source of the quotation "Music is organized sound."

Dennett, Daniel C. 2005. Show me the science. *The New York Times*, August 28.
> Source of the quotation "Heat is not made of tiny hot things."

Doyle, P. 2005. *Echo & Reverb: Fabricating Space in Popular Music Recording, 1900–1960*. Middletown, Conn.

An expansive, scholarly survey of the recording industry's fascination
with space and creating artificial ambiences.

Dwyer, T. 1971. *Composing with Tape Recorders: Musique Concrète*. New York:
Oxford University Press.
For background information on the *musique concrète* of Schaeffer,
Dhomon, Normandeau, and others.

*Grey, J. M. 1975. An exploration of musical timbre using computer-based tech-
niques for analysis, synthesis, and perceptual scaling. Ph.D. Thesis, Music,
Center for Computer Research in Music and Acoustics, Stanford University,
Stanford, Calif.
The most influential paper on modern approaches to the study of timbre.

*Janata, P. 1997. Electrophysiological studies of auditory contexts. Dissertation
Abstracts International: Section B: The Sciences and Engineering, University
of Oregon.
Contains the experiments showing that the inferior colliculus of the barn
owl restores the missing fundamental.

*Krumhansl, C. L. 1990. *Cognitive Foundations of Musical Pitch*. New York: Oxford University Press.

*———. 1991. Music psychology: Tonal structures in perception and memory.
Annual Review of Psychology 42:277–303.

*———. 2000. Rhythm and pitch in music cognition. *Psychological Bulletin* 126
(1):159–179.

*———. 2002. Music: A link between cognition and emotion. *Current Direc-
tions in Psychological Science* 11 (2):45–50.
Krumhansl is one of the leading scientists working in music perception
and cognition; these articles, and the monograph, provide foundations of
the field, and in particular, the notion of tonal hierarchies, the dimension-
ality of pitch, and the mental representation of pitch.

*Kubovy, M. 1981. Integral and separable dimensions and the theory of indispens-
able attributes. In *Perceptual Organization*, edited by M. Kubovy and J. Pomer-
antz. Hillsdale, N.J.: Erlbaum.
Source for the notion of separable dimensions in music.

Levitin, D. J. 2002. Memory for musical attributes. In *Foundations of Cognitive
Psychology: Core Readings*, edited by D. J. Levitin. Cambridge: MIT Press.
Source for the listing of eight different perceptual attributes of a sound.

*McAdams, S., J. W. Beauchamp, and S. Meneguzzi. 1999. Discrimination of mu-
sical instrument sounds resynthesized with simplified spectrotemporal parameters. *Journal of the Acoustical Society of America* 105 (2):882–897.

McAdams, S., and E. Bigand. 1993. Introduction to auditory cognition. In *Thinking in Sound: The Cognitive Psychology of Audition*, edited by S. McAdams and E. Bigand. Oxford: Clarendon Press.

*McAdams, S., and J. Cunible. 1992. Perception of timbral analogies. *Philosophical Transactions of the Royal Society of London*, B 336:383–389.

*McAdams, S., S. Winsberg, S. Donnadieu, and G. De Soete. 1995. Perceptual scaling of synthesized musical timbres: Common dimensions, specificities, and latent subject classes. *Psychological Research/Psychologische Forschung* 58 (3):177–192.

McAdams is the leading researcher in the world studying timbre, and these four papers provide an overview of what we currently know about timbre perception.

Newton, I. 1730/1952. *Opticks: or, A Treatise of the Reflections, Refractions, Inflections, and Colours of Light.* New York: Dover.

Source for Newton's observation that light waves are not themselves colored.

*Oxenham, A. J., J. G. W. Bernstein, and H. Penagos. 2004. Correct tonotopic representation is necessary for complex pitch perception. *Proceedings of the National Academy of Sciences* 101:1421–1425.

On tonotopic representations of pitch in the auditory system.

Palmer, S. E. 2000. *Vision: From Photons to Phenomenology.* Cambridge: MIT Press.

An excellent introduction to cognitive science and vision science, at the undergraduate level. Full disclosure: Palmer and I are collaborators, and I made some contributions to this book. Source for the different attributes of visual stimuli.

Pierce, J. R. 1992. *The Science of Musical Sound*, revised ed. San Francisco: W. H. Freeman.

Excellent source for the educated layperson who wants to understand the physics of sound, overtones, scales, etc. Full disclosure: Pierce was my teacher and friend when he was alive.

Rossing, T. D. 1990. *The Science of Sound*, 2nd ed. Reading, Mass.: Addison-Wesley Publishing.

Another excellent source for the physics of sound, overtones, scales, and so on, appropriate for undergraduates.

Schaeffer, Pierre. 1967. *La musique concrète.* Paris: Presses Universitaires de France.

———. 1968. *Traité des objets musicaux.* Paris: Le Seuil.

The principles of *musique concrète* are introduced in the first work, and Schaeffer's masterpiece on the theory of sound in the second. Unfortunately, no English translation yet exists.

Schmeling, P. 2005. *Berklee Music Theory Book 1*. Boston: Berklee Press.
I learned music theory at Berklee College, and this is the first volume in their set. Suitable for self-teaching, this covers all the basics.

*Schroeder, M. R. 1962. Natural sounding artificial reverberation. *Journal of the Audio Engineering Society* 10 (3):219–233.
The seminal article on the creation of artificial reverberation.

Scorsese, Martin. 2005. *No Direction Home*. USA: Paramount.
Source of the reports of Bob Dylan being booed at the Newport Folk Festival.

Sethares, W. A. 1997. *Tuning, Timbre, Spectrum, Scale*. London: Springer.
A rigorous introduction to the physics of music and musical sounds.

*Shamma, S., and D. Klein. 2000. The case of the missing pitch templates: How harmonic templates emerge in the early auditory system. *Journal of the Acoustical Society of America* 107 (5):2631–2644.

*Shamma, S. A. 2004. Topographic organization is essential for pitch perception. *Proceedings of the National Academy of Sciences* 101:1114–1115.
On tonotopic representations of pitch in the auditory system.

*Smith, J. O., III. 1992. Physical modeling using digital waveguides. *Computer Music Journal* 16 (4):74–91.
The article that introduced wave guide synthesis.

Surmani, A., K. F. Surmani, and M. Manus. 2004. *Essentials of Music Theory: A Complete Self-Study Course for All Musicians*. Van Nuys, Calif.: Alfred Publishing Company.
Another excellent self-teaching guide to music theory.

Taylor, C. 1992. *Exploring Music: The Science and Technology of Tones and Tunes*. Bristol: Institute of Physics Publishing.
Another excellent college-level text on the physics of sound.

Trehhub, S. E. 2003. Musical predispositions in infancy. In *The Cognitive Neuroscience of Music*, edited by I. Perets and R. J. Zatorre. Oxford: Oxford University Press.

*Västfjäll, D., P. Larsson, and M. Kleiner. 2002. Emotional and auditory virtual environments: Affect-based judgments of music reproduced with virtual reverberation times. *CyberPsychology & Behavior* 5 (1):19–32.
A recent scholarly article on the effect of reverberation on emotional response.

Chapter 2

*Bregman, A. S. 1990. *Auditory Scene Analysis*. Cambridge: MIT Press.
The definitive work on general auditory grouping principles.

Clarke, E. F. 1999. Rhythm and timing in music. In *The Psychology of Music*, edited by D. Deutsch. San Diego: Academic Press.
> An undergraduate-level article on the psychology of time perception in music, and the source for the Eric Clarke quote.

*Ehrenfels, C. von. 1890/1988. On "Gestalt qualities." In *Foundations of Gestalt Theory*, edited by B. Smith. Munich: Philosophia Verlag.
> On the founding of Gestalt psychology and the Gestalt psychologists' interest in melody.

Elias, L. J., and D. M. Saucier. 2006. *Neuropsychology: Clinical and Experimental Foundations*. Boston: Pearson.
> Textbook for introducing fundamental concepts of neuroanatomy and the functions of different brain regions.

*Fishman, Y. I., D. H. Reser, J. C. Arezzo, and M. Steinschneider. 2000. Complex tone processing in primary auditory cortex of the awake monkey. I. Neural ensemble correlates of roughness. *Journal of the Acoustical Society of America* 108:235–246.
> The physiological basis of consonance and dissonance perception.

Gilmore, Mikal. 2005. Lennon lives forever: Twenty-five years after his death, his music and message endure. *Rolling Stone*, December 15.
> Source of the John Lennon quote.

Helmholtz, H. L. F. 1885/1954. *On the Sensations of Tone*, 2nd revised ed. New York: Dover.
> Unconscious inference.

Lerdahl, Fred. 1983. *A Generative Theory of Tonal Music*. Cambridge: MIT Press.
> The most influential statement of auditory grouping principles in music.

*Levitin, D. J., and P. R. Cook. 1996. Memory for musical tempo: Additional evidence that auditory memory is absolute. *Perception and Psychophysics* 58:927–935.
> This is the article mentioned in the text, in which Cook and I asked people to sing their favorite rock songs, and they reproduced the tempo with very high accuracy.

Luce, R. D. 1993. *Sound and Hearing: A Conceptual Introduction*. Hillsdale, N.J.: Erlbaum.
> Textbook on the ear and hearing, including physiology of the ear, loudness, pitch perception, etc.

*Mesulam, M.-M. 1985. *Principles of Behavioral Neurology*. Philadelphia: F. A. Davis Company.
> Advanced, graduate textbook for introducing fundamental concepts of neuroanatomy and the functions of different brain regions.

Moore, B. C. J. 1982. *An Introduction to the Psychology of Hearing*, 2nd ed. London: Academic Press.

————. 2003. *An Introduction to the Psychology of Hearing*, 5th ed. Amsterdam: Academic Press.
> Textbooks on the ear and hearing, including physiology of the ear, loudness, pitch perception, etc.

Palmer, S. E. 2002. Organizing objects and scenes. In *Foundations of Cognitive Psychology: Core readings*, edited by D. J. Levitin. Cambridge: MIT Press.
> On the Gestalt principles of visual grouping.

Stevens, S. S., and F. Warshofsky. 1965. *Sound and Hearing*, edited by R. Dubos, H. Margenau, C. P. Snow. *Life* Science Library. New York: Time Incorporated.
> A good introduction to the principles of hearing and auditory perception for the general reader.

*Tramo, M. J., P. A. Cariani, B. Delgutte, and L. D. Braida. 2003. Neurobiology of harmony perception. In *The Cognitive Neuroscience of Music*, edited by I. Peretz and R. J. Zatorre. New York: Oxford University Press.
> The physiological basis of consonance and dissonance perception.

Yost, W. A. 1994. *Fundamentals of Hearing: An Introduction*, 3rd ed. San Diego: Academic Press, Inc.
> Textbook on hearing, pitch, and loudness perception.

Zimbardo, P. G., and R. J. Gerrig. 2002. Perception. In *Foundations of Cognitive Psychology*, edited by D. J. Levitin. Cambridge: MIT Press.
> The Gestalt principles of grouping.

Chapter 3

Bregman, A. S. 1990. *Auditory Scene Analysis*. Cambridge: MIT Press.
> Streaming by timbre and other auditory grouping principles. My analogy about the eardrum as a pillowcase stretched over a bucket borrows liberally from a different analogy Bregman proposes in this book.

*Chomsky, N. 1957. *Syntactic Structures*. The Hague, Netherlands: Mouton.
> About the innateness of a language capacity in the human brain.

Crick, F. H. C. 1995. *The Astonishing Hypothesis: The Scientific Search for the Soul*. New York: Touchstone/Simon & Schuster.
> The idea that all of human behavior can be explained by the activity of the brain and neurons.

Dennett, D. C. 1991. *Consciousness Explained*. Boston: Little, Brown and Company.
> On the illusions of conscious experience, and brains updating information.

————. 2002. Can machines think? In *Foundations of Cognitive Psychology: Core Readings*, edited by D. J. Levitin. Cambridge: MIT Press.

————. 2002. Where am I? In *Foundations of Cognitive Psychology: Core Readings*, edited by D. J. Levitin. Cambridge: MIT Press.

These two articles address foundational issues of the brain as computer and the philosophical idea of *functionalism;* "Can Machines Think?" also summarizes the Turing test for intelligence, and its strengths and weaknesses.

*Friston, K. J. 2005. Models of brain function in neuroimaging. *Annual Review of Psychology* 56:57–87.
    A technical overview on research methods for the analysis of brain imaging data by one of the inventors of SPM, a widely used statistical package for fMRI data.

Gazzaniga, M. S., R. B. Ivry, and G. Mangun. 1998. *Cognitive Neuroscience*. New York: Norton.
    Functional divisions of the brain; basic divisions into lobes, major anatomical landmarks; undergraduate text.

Gertz, S. D., and R. Tadmor. 1996. *Liebman's Neuroanatomy Made Easy and Understandable*, 5th ed. Gaithersburg, Md.: Aspen.
    An introduction to neuroanatomy and major brain regions.

Gregory, R. L. 1986. *Odd Perceptions*. London: Routledge.
    On perception as inference.

*Griffiths, T. D., S. Uppenkamp, I. Johnsrude, O. Josephs, and R. D. Patterson. 2001. Encoding of the temporal regularity of sound in the human brainstem. *Nature Neuroscience* 4 (6):633–637.

*Griffiths, T. D., and J. D. Warren. 2002. The planum temporale as a computational hub. *Trends in Neuroscience* 25 (7):348–353.
    Recent work on sound processing in the brain from Griffiths, one of the most esteemed researchers of the current generation of brain scientists studying auditory processes.

*Hickok, G., B. Buchsbaum, C. Humphries, and T. Muftuler. 2003. Auditory-motor interaction revealed by fMRI: Speech, music, and working memory in area Spt. *Journal of Cognitive Neuroscience* 15 (5):673–682.
    A primary source for music activation in a brain region at the posterior Sylvian fissure at the parietal-temporal boundary.

*Janata, P., J. L. Birk, J. D. Van Horn, M. Leman, B. Tillmann, and J. J. Bharucha. 2002. The cortical topography of tonal structures underlying Western music. *Science* 298:2167–2170.

*Janata, P., and S. T. Grafton. 2003. Swinging in the brain: Shared neural substrates for behaviors related to sequencing and music. *Nature Neuroscience* 6 (7):682–687.

*Johnsrude, I. S., V. B. Penhune, and R. J. Zatorre. 2000. Functional specificity in the right human auditory cortex for perceiving pitch direction. *Brain Res Cogn Brain Res* 123:155–163.

*Knosche, T. R., C. Neuhaus, J. Haueisen, K. Alter, B. Maess, O. Witte, and A. D. Friederici. 2005. Perception of phrase structure in music. *Human Brain Mapping* 24 (4):259–273.

*Koelsch, S., E. Kasper, D. Sammler, K. Schulze, T. Gunter, and A. D. Friederici. 2004. Music, language and meaning: brain signatures of semantic processing. *Nature Neuroscience* 7 (3):302–307.

*Koelsch, S., E. Schröger, and T. C. Gunter. 2002. Music matters: Preattentive musicality of the human brain. *Psychophysiology* 39 (1):38–48.

*Kuriki, S., N. Isahai, T. Hasimoto, F. Takeuchi, and Y. Hirata. 2000. Music and language: Brain activities in processing melody and words. Paper read at 12th International Conference on Biomagnetism.
> Primary sources on the neuroanatomy of music perception and cognition.

Levitin, D. J. 1996. High-fidelity music: Imagine listening from inside the guitar. *The New York Times*, December 15.

———. 1996. The modern art of studio recording. *Audio*, September, 46–52.
> On modern recording techniques and the illusions they create.

———. 2002. Experimental design in psychological research. In *Foundations of Cognitive Psychology: Core Readings*, edited by D. J. Levitin. Cambridge: MIT Press.
> On experimental design and what is a "good" experiment.

*Levitin, D. J., and V. Menon. 2003. Musical structure is processed in "language" areas of the brain: A possible role for Brodmann Area 47 in temporal coherence. *NeuroImage* 20 (4):2142–2152.
> The first research article using fMRI to show that temporal structure and temporal coherence in music is processed in the same brain region that does so for spoken and signed languages.

*McClelland, J. L., D. E. Rumelhart, and G. E. Hinton. 2002. The appeal of parallel distributed processing. In *Foundations of Cognitive Psychology: Core Readings*, edited by D. J. Levitin. Cambridge: MIT Press.
> The brain as a parallel processing machine.

Palmer, S. 2002. Visual awareness. In *Foundations of Cognitive Psychology: Core Readings*, edited by D. J. Levitin. Cambridge: MIT Press.
> The philosophical foundations of modern cognitive science, dualism, and materialism.

*Parsons, L. M. 2001. Exploring the functional neuroanatomy of music performance, perception, and comprehension. In I. Peretz and R. J. Zatorre, Eds., *Biological Foundations of Music*, Annals of the New York Academy of Sciences, Vol. 930, pp. 211–230.

*Patel, A. D., and E. Balaban. 2004. Human auditory cortical dynamics during perception of long acoustic sequences: Phase tracking of carrier frequency by the auditory steady-state response. *Cerebral Cortex* 14 (1):35–46.

*Patel, A. D. 2003. Language, music, syntax, and the brain. *Nature Neuroscience* 6 (7):674–681.

*Patel, A. D., and E. Balaban. 2000. Temporal patterns of human cortical activity reflect tone sequence structure. *Nature* 404:80–84.

*Peretz, I. 2000. Music cognition in the brain of the majority: Autonomy and fractionation of the music recognition system. In *The Handbook of Cognitive Neuropsychology*, edited by B. Rapp. Hove, U.K.: Psychology Press.

*Peretz, I. 2000. Music perception and recognition. In *The Handbook of Cognitive Neuropsychology*, edited by B. Rapp. Hove, U.K.: Psychology Press.

*Peretz, I., and M. Coltheart. 2003. Modularity of music processing. *Nature Neuroscience* 6 (7):688–691.

*Peretz, I., and L. Gagnon. 1999. Dissociation between recognition and emotional judgements for melodies. *Neurocase* 5:21–30.

*Peretz, I., and R. J. Zatorre, eds. 2003. *The Cognitive Neuroscience of Music*. New York: Oxford.
    Primary sources on the neuroanatomy of music perception and cognition.

Pinker, S. 1997. *How The Mind Works*. New York: W. W. Norton.
    Pinker claims here that music is an evolutionary accident.

*Posner, M. I. 1980. Orienting of attention. *Quarterly Journal of Experimental Psychology* 32:3–25.
    The Posner Cueing Paradigm.

Posner, M. I., and D. J. Levitin. 1997. Imaging the future. In *The Science of the Mind: The 21st Century*. Cambridge: MIT Press.
    A more complete explanation of the bias that Posner and I have against simple "mental cartography" for its own sake.

Ramachandran, V. S. 2004. *A Brief Tour of Human Consciousness: From Impostor Poodles to Purple Numbers*. New York: Pi Press.
    Consciousness and our naive intuitions about it.

*Rock, I. 1983. *The Logic of Perception*. Cambridge: MIT Press.
    Perception as a logical process and as constructive.

*Schmahmann, J. D., ed. 1997. *The Cerebellum and Cognition*. San Diego: Academic Press.
    On the cerebellum's role in emotional regulation.

Searle, J. R. 2002. Minds, brains, and programs. In *Foundations of Cognitive Psychology: Core Readings*, edited by D. J. Levitin. Cambridge: MIT Press.
    The brain as a computer; this is one of the most discussed, argued, and cited articles in modern philosophy of mind.

*Sergent, J. 1993. Mapping the musician brain. *Human Brain Mapping* 1:20–38.
One of the first neuroimaging reports of music and the brain, still widely cited and referred to.

Shepard, R. N. 1990. *Mind Sights: Original Visual Illusions, Ambiguities, and Other Anomalies, with a Commentary on the Play of Mind in Perception and Art*. New York: W. H. Freeman.
Source of the "Turning the Tables" illusion.

*Steinke, W. R., and L. L. Cuddy. 2001. Dissociations among functional subsystems governing melody recognition after right hemisphere damage. *Cognitive Neuroscience* 18 (5):411–437.

*Tillmann, B., P. Janata, and J. J. Bharucha. 2003. Activation of the inferior frontal cortex in musical priming. *Cognitive Brain Research* 16:145–161.
Primary sources on the neuroanatomy of music perception and cognition.

*Warren, R. M. 1970. Perceptual restoration of missing speech sounds. *Science*, January 23, 392–393.
Source of the example of auditory "filling in" or perceptual completion.

Weinberger, N. M. 2004. Music and the Brain. *Scientific American* (November 2004):89–95.

*Zatorre, R. J., and P. Belin. 2001. Spectral and temporal processing in human auditory cortex. *Cerebral Cortex* 11:946–953.

*Zatorre, R. J., P. Belin, and V. B. Penhune. 2002. Structure and function of auditory cortex: Music and speech. *Trends in Cognitive Sciences* 6 (1):37–46.
Primary sources on the neuroanatomy of music perception and cognition.

Chapter 4

*Bartlett, F. C. 1932. *Remembering: A Study in Experimental and Social Psychology*. London: Cambridge University Press.
On schemas.

*Bavelier, D., C. Brozinsky, A. Tomann, T. Mitchell, H. Neville, and G. Liu. 2001. Impact of early deafness and early exposure to sign language on the cerebral organization for motion processing. *The Journal of Neuroscience* 21 (22):8931–8942.

*Bavelier, D., D. P. Corina, and H. J. Neville. 1998. Brain and language: A perspective from sign language. *Neuron* 21:275–278.
The neuroanatomy of sign language.

*Bever, T. G., and Chiarell, R. J. 1974. Cerebral dominance in musicians and nonmusicians. *Science* 185 (4150):537–539.
A seminal paper on hemispheric specialization for music.

*Bharucha, J. J. 1987. Music cognition and perceptual facilitation—a connectionist framework. *Music Perception* 5 (1):1–30.

*———. 1991. Pitch, harmony, and neural nets: A psychological perspective. In *Music and Connectionism*, edited by P. M. Todd and D. G. Loy. Cambridge: MIT Press.

*Bharucha, J. J., and P. M. Todd. 1989. Modeling the perception of tonal structure with neural nets. *Computer Music Journal* 13 (4):44–53.

*Bharucha, J. J. 1992. Tonality and learnability. In *Cognitive Bases of Musical Communication*, edited by M. R. Jones and S. Holleran. Washington, D.C: American Psychological Association.
    On musical schemas.

*Binder, J., and C. J. Price. 2001. Functional neuroimaging of language. In *Handbook of Functional Neuroimaging of Cognition*, edited by A. Cabeza and A. Kingston.

*Binder, J. R., E. Liebenthal, E. T. Possing, D. A. Medler, and B. D. Ward. 2004. Neural correlates of sensory and decision processes in auditory object identification. *Nature Neuroscience* 7 (3):295–301.

*Bookheimer, S. Y. 2002. Functional MRI of language: New approaches to understanding the cortical organization of semantic processing. *Annual Review of Neuroscience* 25:151–188.
    The functional neuroanatomy of speech.

Cook, P. R. 2005. The deceptive cadence as a parlor trick. Princeton, N.J., Montreal, Que., November 30.
    Personal communication from Perry Cook, who described the deceptive cadence this way in an e-mail to me.

*Cowan, W. M., T. C. Südhof, and C. F. Stevens, eds. 2001. *Synapses*. Baltimore: Johns Hopkins University Press.
    In-depth information on synapses, the synaptic cleft, and synaptic transmission.

*Dibben, N. 1999. The perception of structural stability in atonal music: the influence of salience, stability, horizontal motion, pitch commonality, and dissonance. *Music Perception* 16 (3):265–24.
    On atonal music, such as that by Schönberg described in this chapter.

*Franceries, X., B. Doyon, N. Chauveau, B. Rigaud, P. Celsis, and J.-P. Morucci. 2003. Solution of Poisson's equation in a volume conductor using resistor mesh models: Application to event related potential imaging. *Journal of Applied Physics* 93 (6):3578–3588.
    The inverse Poisson problem of localization with EEG.

Fromkin, V., and R. Rodman. 1993. *An Introduction to Language*, 5th ed. Fort Worth, Tex.: Harcourt Brace Jovanovich College Publishers.
    The basics of psycholinguistics, phonemes, word formation.

*Gazzaniga, M. S. 2000. *The New Cognitive Neurosciences*, 2nd ed. Cambridge: MIT Press.
    Foundations of neuroscience.

Gernsbacher, M. A., and M. P. Kaschak. 2003. Neuroimaging studies of language production and comprehension. *Annual Review of Psychology* 54:91–114.
    A recent review of studies of the neuroanatomical basis for language.

*Hickok, G., B. Buchsbaum, C. Humphries, and T. Muftuler. 2003. Auditory-motor interaction revealed by fMRI: Speech, music, and working memory in area Spt. *Journal of Cognitive Neuroscience* 15 (5):673–682.

*Hickok, G., and Poeppel, D. 2000. Towards a functional neuroanatomy of speech perception. *Trends in Cognitive Sciences* 4 (4):131–138.
    The neuroanatomical basis for speech and music.

Holland, B. 1981. A man who sees what others hear. *The New York Times*, November 19.
    An article about Arthur Lintgen, the man who can read record grooves. He can only read them for music that he knows, and only for classical music post-Beethoven.

*Huettel, S. A., A. W. Song, and G. McCarthy. 2003. *Functional Magnetic Resonance Imaging*. Sunderland, Mass.: Sinauer Associates, Inc.
    On the theory behind fMRI.

*Ivry, R. B., and L. C. Robertson. 1997. *The Two Sides of Perception*. Cambridge: MIT Press.
    On hemispheric specialization.

*Johnsrude, I. S., V. B. Penhune, and R. J. Zatorre. 2000. Functional specificity in the right human auditory cortex for perceiving pitch direction. *Brain Res Cogn Brain Res* 123:155–163.

*Johnsrude, I. S., R. J. Zatorre, B. A. Milner, and A. C. Evans. 1997. Left-hemisphere specialization for the processing of acoustic transients. *NeuroReport* 8:1761–1765.
    The neuroanatomy of speech and music.

*Kandel, E. R., J. H. Schwartz, and T. M. Jessell. 2000. *Principles of Neural Science*, 4th ed. New York: McGraw-Hill.
    Foundations of neuroscience, cowritten by Nobel Laureate Eric Kandel. This is a widely used text in medical schools and graduate neuroscience programs.

*Knosche, T. R., C. Neuhaus, J. Haueisen, K. Alter, B. Maess, O. Witte, and A. D. Friederici. 2005. Perception of phrase structure in music. *Human Brain Mapping* 24 (4):259–273.

*Koelsch, S., T. C. Gunter, D. Y. v. Cramon, S. Zysset, G. Lohmann, and A. D. Friederici. 2002. Bach speaks: A cortical "language-network" serves the processing of music. *NeuroImage* 17:956–966.

*Koelsch, S., E. Kasper, D. Sammler, K. Schulze, T. Gunter, and A. D. Friederici. 2004. Music, language, and meaning: Brain signatures of semantic processing. *Nature Neuroscience* 7 (3):302–307.

*Koelsch, S., B. Maess, and A. D. Friederici. 2000. Musical syntax is processed in the area of Broca: an MEG study. *NeuroImage* 11 (5):56.
    Articles on musical structure by Koelsch, Friederici, and their colleagues.

Kosslyn, S. M., and O. Koenig. 1992. *Wet Mind: The New Cognitive Neuroscience*. New York: Free Press.
    A general audience's introduction to cognitive neuroscience.

*Krumhansl, C. L. 1990. *Cognitive Foundations of Musical Pitch*. New York: Oxford University Press.
    On the dimensionality of pitch.

*Lerdahl, F. 1989. Atonal prolongational structure. *Contemporary Music Review* 3 (2).
    On atonal music, such as that of Schönberg.

*Levitin, D. J., and V. Menon. 2003. Musical structure is processed in "language" areas of the brain: A possible role for Brodmann Area 47 in temporal coherence. *NeuroImage* 20 (4):2142–2152.

*———. 2005. The neural locus of temporal structure and expectancies in music: Evidence from functional neuroimaging at 3 Tesla. *Music Perception* 22 (3):563–575.
    The neuroanatomy of musical structure.

*Maess, B., S. Koelsch, T. C. Gunter, and A. D. Friederici. 2001. Musical syntax is processed in Broca's area: An MEG study. *Nature Neuroscience* 4 (5):540–545.
    The neuroanatomy of musical structure.

*Marin, O. S. M. 1982. Neurological aspects of music perception and performance. In *The Psychology of Music*, edited by D. Deutsch. New York: Academic Press.
    Loss of musical function due to lesions.

*Martin, R. C. 2003. Language processing: Functional organization and neuroanatomical basis. *Annual Review of Psychology* 54:55–89.
    The neuroanatomy of speech perception.

McClelland, J. L., D. E. Rumelhart, and G. E. Hinton. 2002. The Appeal of Parallel Distributed Processing. In *Foundations of Cognitive Psychology: Core Readings*, edited by D. J. Levitin. Cambridge: MIT Press.
    On schemas.

Meyer, L. B. 2001. Music and emotion: distinctions and uncertainties. In *Music and Emotion: Theory and Research*, edited by P. N. Juslin and J. A. Sloboda. Oxford and New York: Oxford University Press.

Meyer, Leonard B. 1956. *Emotion and Meaning in Music*. Chicago: University of Chicago Press.

————. 1994. *Music, the Arts, and Ideas: Patterns and Predictions in Twentieth-Century Culture*. Chicago: University of Chicago Press.
      On musical style, repetition, gap-fill, and expectations.

*Milner, B. 1962. Laterality effects in audition. In *Interhemispheric Effects and Cerebral Dominance*, edited by V. Mountcastle. Baltimore: Johns Hopkins Press.
      Laterality in hearing.

*Narmour, E. 1992. *The Analysis and Cognition of Melodic Complexity: The Implication-Realization Model*. Chicago: University of Chicago Press.

*————. 1999. Hierarchical expectation and musical style. In *The Psychology of Music*, edited by D. Deutsch. San Diego: Academic Press.
      On musical style, repetition, gap-fill, and expectations.

*Niedermeyer, E., and F. L. Da Silva. 2005. *Electroencephalography: Basic Principles, Clinical Applications, and Related Fields*, 5th ed. Philadephia: Lippincott, Williams & Wilkins.
      An introduction to EEG (advanced, technical, not for the faint of heart).

*Panksepp, J., ed. 2002. *Textbook of Biological Psychiatry*. Hoboken, N.J.: Wiley.
      On SSRIs, seratonin, dopamine, and neurochemistry.

*Patel, A. D. 2003. Language, music, syntax and the brain. *Nature Neuroscience* 6 (7):674–681.
      The neuroanatomy of musical structure; this paper introduces the SSIRH.

*Penhune, V. B., R. J. Zatorre, J. D. MacDonald, and A. C. Evans. 1996. Interhemispheric anatomical differences in human primary auditory cortex: Probabilistic mapping and volume measurement from magnetic resonance scans. *Cerebral Cortex* 6:661–672.

*Peretz, I., R. Kolinsky, M. J. Tramo, R. Labrecque, C. Hublet, G. Demeurisse, and S. Belleville. 1994. Functional dissociations following bilateral lesions of auditory cortex. *Brain* 117:1283–1301.

*Perry, D. W., R. J. Zatorre, M. Petrides, B. Alivisatos, E. Meyer, and A. C. Evans. 1999. Localization of cerebral activity during simple singing. *NeuroReport* 10:3979–3984.
      The neuroanatomy of music processing.

*Petitto, L. A., R. J. Zatorre, K. Gauna, E. J. Nikelski, D. Dostie, and A. C. Evans. 2000. Speech-like cerebral activity in profoundly deaf people processing signed

languages: Implications for the neural basis of human language. *Proceedings of the National Academy of Sciences* 97 (25):13961–13966.
>The neuroanatomy of sign language.

Posner, M. I. 1973. *Cognition: An Introduction*. Edited by J. L. E. Bourne and L. Berkowitz, 1st ed. Basic Psychological Concepts Series. Glenview, Ill.: Scott, Foresman and Company.

———. 1986. *Chronometric Explorations of Mind: The Third Paul M. Fitts Lectures, Delivered at the University of Michigan, September 1976.* New York: Oxford University Press.
>On mental codes.

Posner, M. I., and M. E. Raichle. 1994. *Images of Mind.* New York: Scientific American Library.
>A general-reader introduction to neuroimaging.

Rosen, C. 1975. *Arnold Schoenberg.* Chicago: University of Chicago Press.
>On the composer, atonal and twelve-tone music.

*Russell, G. S., K. J. Eriksen, P. Poolman, P. Luu, and D. Tucker. 2005. Geodesic photogrammetry for localizing sensor positions in dense-array EEG. *Clinical Neuropsychology* 116:1130–1140.
>The inverse Poisson problem in EEG localization.

Samson, S., and R. J. Zatorre. 1991. Recognition memory for text and melody of songs after unilateral temporal lobe lesion: Evidence for dual encoding. *Journal of Experimental Psychology: Learning, Memory, and Cognition* 17 (4):793–804.

———. 1994. Contribution of the right temporal lobe to musical timbre discrimination. *Neuropsychologia* 32:231–240.
>Neuroanatomy of music and speech perception.

Schank, R. C., and R. P. Abelson. 1977. *Scripts, plans, goals, and understanding.* Hillsdale, N.J.: Lawrence Erlbaum Associates.
>Seminal work on schemas.

*Shepard, R. N. 1964. Circularity in judgments of relative pitch. *Journal of The Acoustical Society of America* 36 (12):2346–2353.

*———. 1982. Geometrical approximations to the structure of musical pitch. *Psychological Review* 89 (4):305–333.

*———. 1982. Structural representations of musical pitch. In *Psychology of Music*, edited by D. Deutsch. San Diego: Academic Press.
>The dimensionality of pitch.

Squire, L. R., F. E. Bloom, S. K. McConnell, J. L. Roberts, N. C. Spitzer, and M. J. Zigmond, eds. 2003. *Fundamental Neuroscience*, 2nd ed. San Diego: Academic Press.
>Basic neuroscience text.

*Temple, E., R. A. Poldrack, A. Protopapas, S. S. Nagarajan, T. Salz, P. Tallal, M. M. Merzenich, and J. D. E. Gabrieli. 2000. Disruption of the neural response to rapid acoustic stimuli in dyslexia: Evidence from functional MRI. *Proceedings of the National Academy of Sciences* 97 (25):13907–13912.
    Functional neuroanatomy of speech.

*Tramo, M. J., J. J. Bharucha, and F. E. Musiek. 1990. Music perception and cognition following bilateral lesions of auditory cortex. *Journal of Cognitive Neuroscience* 2:195–212.

*Zatorre, R. J. 1985. Discrimination and recognition of tonal melodies after unilateral cerebral excisions. *Neuropsychologia* 23 (1):31–41.

*———. 1998. Functional specialization of human auditory cortex for musical processing. *Brain* 121 (Part 10):1817–1818.

*Zatorre, R. J., P. Belin, and V. B. Penhune. 2002. Structure and function of auditory cortex: Music and speech. *Trends in Cognitive Sciences* 6 (1):37–46.

*Zatorre, R. J., A. C. Evans, E. Meyer, and A. Gjedde. 1992. Lateralization of phonetic and pitch discrimination in speech processing. *Science* 256 (5058):846–849.

*Zatorre, R. J., and S. Samson. 1991. Role of the right temporal neocortex in retention of pitch in auditory short-term memory. *Brain* (114):2403–2417.
    Studies of the neuroanatomy of speech and music, and of the effect of lesions.

Chapter 5

Bjork, E. L., and R. A. Bjork, eds. 1996. *Memory, Handbook of Perception and Cognition*, 2nd ed. San Diego: Academic Press.
    General text on memory for the researcher.

Cook, P. R., ed. 1999. *Music, Cognition, and Computerized Sound: An Introduction to Psychoacoustics*. Cambridge: MIT Press.
    This book consists of the lectures that I attended as an undergraduate in the course I mention, taught by Pierce, Chowning, Mathews, Shepard, and others.

*Dannenberg, R. B., B. Thom, and D. Watson. 1997. A machine learning approach to musical style recognition. Paper read at International Computer Music Conference, September. Thessoloniki, Greece.
    A source article about music fingerprinting.

Dowling, W. J., and D. L. Harwood. 1986. *Music Cognition*. San Diego: Academic Press.
    On the recognition of melodies in spite of transformations.

Gazzaniga, M. S., R. B. Ivry, and G. R. Mangun. 1998. *Cognitive Neuroscience: The Biology of the Mind*. New York: W. W. Norton.
    Contains a summary of Gazzaniga's split-brain studies.

*Goldinger, S. D. 1996. Words and voices: Episodic traces in spoken word identification and recognition memory. *Journal of Experimental Psychology: Learning, Memory, and Cognition* 22 (5):1166–1183.

*———. 1998. Echoes of echoes? An episodic theory of lexical access. *Psychological Review* 105 (2):251–279.
  Source articles on multiple-trace memory theory.

Guenther, R. K. 2002. Memory. In *Foundations of Cognitive Psychology: Core Readings*, edited by D. J. Levitin. Cambridge: MIT Press.
  An overview of the record-keeping vs. constructivist theories of memory.

*Haitsma, J., and T. Kalker. 2003. A highly robust audio fingerprinting system with an efficient search strategy. *Journal of New Music Research* 32 (2):211–221.
  Another source article on audio fingerprinting.

*Halpern, A. R. 1988. Mental scanning in auditory imagery for songs. *Journal of Experimental Psychology: Learning, Memory, and Cognition* 143:434–443.
  Source for the discussion in this chapter about the ability to scan music in our heads.

*———. 1989. Memory for the absolute pitch of familiar songs. *Memory and Cognition* 17 (5):572–581.
  This article was the inspiration for my 1994 study.

*Heider, E. R. 1972. Universals in color naming and memory. *Journal of Experimental Psychology* 93 (1):10–20.
  Under Eleanor Rosch's married name, a foundational work on categorization.

*Hintzman, D. H. 1986. "Schema abstraction" in a multiple-trace memory model. *Psychological Review* 93 (4):411–428.
  Hintzman's MINERVA model is discussed in the context of multiple-trace memory models.

*Hintzman, D. L., R. A. Block, and N. R. Inskeep. 1972. Memory for mode of input. *Journal of Verbal Learning and Verbal Behavior* 11:741–749.
  Source for the study of fonts that I discuss.

*Ishai, A., L. G. Ungerleider, and J. V. Haxby. 2000. Distributed neural systems for the generation of visual images. *Neuron* 28:979–990.
  Source for the work on categorical separation in the brain.

*Janata, P. 1997. Electrophysiological studies of auditory contexts. Dissertation Abstracts International: Section B: The Sciences and Engineering, University of Oregon.
  This contains the report of imagining a piece of music bearing a nearly identical EEG signature to actually hearing a piece of music.

*Levitin, D. J. 1994. Absolute memory for musical pitch: Evidence from the production of learned melodies. *Perception and Psychophysics* 56 (4):414–423.
>   This is the source article reporting my study of people singing their favorite rock and pop songs at or near the correct key.

*————. 1999. Absolute pitch: Self-reference and human memory. *International Journal of Computing Anticipatory Systems.*
>   An overview of absolute-pitch research.

*————. 1999. Memory for musical attributes. In *Music, Cognition and Computerized Sound: An Introduction to Psychoacoustics,* edited by P. R. Cook. Cambridge: MIT Press.
>   Description of my study with tuning forks and memory for pitch.

————. 2001. Paul Simon: The Grammy interview. *Grammy,* September, 42–46.
>   Source of the Paul Simon comment about listening for timbres.

*Levitin, D. J., and P. R. Cook. 1996. Memory for musical tempo: Additional evidence that auditory memory is absolute. *Perception and Psychophysics* 58:927–935.
>   Source of my study on memory for the tempo of a song.

*Levitin, D. J., and S. E. Rogers. 2005. Pitch perception: Coding, categories, and controversies. *Trends in Cognitive Sciences* 9 (1):26–33.
>   Review of absolute-pitch research.

*Levitin, D. J., and R. J. Zatorre. 2003. On the nature of early training and absolute pitch: A reply to Brown, Sachs, Cammuso and Foldstein. *Music Perception* 21 (1):105–110.
>   A technical note about problems with absolute-pitch research.

Loftus, E. 1979/1996. *Eyewitness Testimony.* Cambridge: Harvard University Press.
>   Source of the experiments on memory distortions.

Luria, A. R. 1968. *The Mind of a Mnemonist.* New York: Basic Books.
>   Source of the story about the patient with hypermnesia.

McClelland, J. L., D. E. Rumelhart, and G. E. Hinton. 2002. The appeal of parallel distributed processing. In *Foundations of Cognitive Psychology: Core Readings,* edited by D. J. Levitin. Cambridge: MIT Press.
>   Seminal article on parallel distributed processing (PDP) models, otherwise known as "neural networks," computer simulations of brain activity.

*McNab, R. J., L. A. Smith, I. H. Witten, C. L. Henderson, and S. J. Cunningham. 1996. Towards the digital music library: tune retrieval from acoustic input. *Proceedings of the First ACM International Conference on Digital Libraries*:11–18.
>   Music fingerprinting overview.

*Parkin, A. J. 1993. *Memory: Phenomena, Experiment and Theory.* Oxford, UK: Blackwell.
>   Textbook on memory.

*Peretz, I., and R. J. Zatorre. 2005. Brain organization for music processing. *Annual Review of Psychology* 56:89–114.
>Review of neuroanatomical foundations of music perception.

*Pope, S. T., F. Holm, and A. Kouznetsov. 2004. Feature extraction and database design for music software. Paper read at International Computer Music Conference in Miami.
>On music fingerprinting.

*Posner, M. I., and S. W. Keele. 1968. On the genesis of abstract ideas. *Journal of Experimental Psychology* 77:353–363.

*———. 1970. Retention of abstract ideas. *Journal of Experimental Psychology* 83:304–308.
>Source for the experiments described that showed prototypes might be stored in memory.

*Rosch, E. 1977. Human categorization. In *Advances in Crosscultural Psychology*, edited by N. Warren. London: Academic Press.

*———. 1978. Principles of categorization. In *Cognition and Categorization*, edited by E. Rosch and B. B. Lloyd. Hillsdale, N.J.: Erlbaum.

*Rosch, E., and C. B. Mervis. 1975. Family resemblances: Studies in the internal structure of categories. *Cognitive Psychology* 7:573–605.

*Rosch, E., C. B. Mervis, W. D. Gray, D. M. Johnson, and P. Boyes-Braem. 1976. Basic objects in natural categories. *Cognitive Psychology* 8:382–439.
>Source articles on Rosch's *prototype theory*.

*Schellenberg, E. G., P. Iverson, and M. C. McKinnon. 1999. Name that tune: Identifying familiar recordings from brief excerpts. *Psychonomic Bulletin & Review* 6 (4):641–646.
>Source for the study described of people naming songs based on timbral cues.

Smith, E. E., and D. L. Medin. 1981. *Categories and concepts*. Cambridge: Harvard University Press.

Smith, E., and D. L. Medin. 2002. The exemplar view. In *Foundations of Cognitive Psychology: Core Readings*, edited by D. J. Levitin. Cambridge: MIT Press.
>On the exemplar view, as an alternative to Rosch's prototype theory.

*Squire, L. R. 1987. *Memory and Brain*. New York: Oxford University Press.
>Textbook on memory.

*Takeuchi, A. H., and S. H. Hulse. 1993. Absolute pitch. *Psychological Bulletin* 113 (2):345–361.

*Ward, W. D. 1999. Absolute Pitch. In *The Psychology of Music*, edited by D. Deutsch. San Diego: Academic Press.
>Overviews of absolute pitch.

*White, B. W. 1960. Recognition of distorted melodies. *American Journal of Psychology* 73:100–107.
  Source for the experiments on how music can be recognized under transposition and other transformations.

Wittgenstein, L. 1953. *Philosophical Investigations*. New York: Macmillan.
  Source for Wittgenstein's writings about "What is a game?" and family resemblance.

Chapter 6

*Desain, P., and H. Honing. 1999. Computational models of beat induction: The rule-based approach. *Journal of New Music Research* 28 (1):29–42.
  This paper discusses some of the algorithms the authors used in the foot-tapping show I wrote about.

*Aitkin, L. M., and J. Boyd. 1978. Acoustic input to lateral pontine nuclei. *Hearing Research* 1 (1):67–77.
  Physiology of the auditory pathway, low-level.

*Barnes, R., and M. R. Jones. 2000. Expectancy, attention, and time. *Cognitive Psychology* 41 (3):254–311.
  An example of Mari Reiss Jones's work on time and timing in music.

Crick, F. 1988. *What Mad Pursuit: A Personal View of Scientific Discovery*. New York: Basic Books.
  Source for the quote about Crick's early years as a scientist.

Crick, F. H. C. 1995. *The Astonishing Hypothesis: The Scientific Search for the Soul*. New York: Touchstone/Simon & Schuster.
  Source for Crick's discussion of reductionism.

*Friston, K. J. 1994. Functional and effective connectivity in neuroimaging: a synthesis. *Human Brain Mapping* 2:56–68.
  The article on functional connectivity that helped Menon to create the analyses we needed for our paper on musical emotion and the nucleus accumbens.

*Gallistel, C. R. 1989. *The Organization of Learning*. Cambridge: MIT Press.
  An example of Randy Gallistel's work.

*Goldstein, A. 1980. Thrills in response to music and other stimuli. *Physiological Psychology* 8 (1):126–129.
  The study that showed that naloxone can block musical emotion.

*Grabow, J. D., M. J. Ebersold, and J. W. Albers. 1975. Summated auditory evoked potentials in cerebellum and inferior colliculus in young rat. *Mayo Clinic Proceedings* 50 (2):57–68.
  Physiology and connections of the cerebellum.

*Holinger, D. P., U. Bellugi, D. L. Mills, J. R. Korenberg, A. L. Reiss, G. F. Sherman, and A. M. Galaburda. In press. Relative sparing of primary auditory cortex in Williams syndrome. *Brain Research.*
> The article that Ursula told Crick about.

*Hopfield, J. J. 1982. Neural networks and physical systems with emergent collective computational abilities. *Proceedings of National Academy of Sciences* 79 (8):2554–2558.
> The first statement of Hopfield nets, a form of neural network model.

*Huang, C., and G. Liu. 1990. Organization of the auditory area in the posterior cerebellar vermis of the cat. *Experimental Brain Research* 81 (2):377–383.

*Huang, C.-M., G. Liu, and R. Huang. 1982. Projections from the cochlear nucleus to the cerebellum. *Brain Research* 244:1–8.

*Ivry, R. B., and R. E. Hazeltine. 1995. Perception and production of temporal intervals across a range of durations: Evidence for a common timing mechanism. *Journal of Experimental Psychology: Human Perception and Performance* 21 (1):3–18.
> Papers on the physiology, anatomy, and connectivity of the cerebellum and lower auditory areas.

*Jastreboff, P. J. 1981. Cerebellar interaction with the acoustic reflex. *Acta Neurobiologiae Experimentalis* 41 (3):279–298.
> Source for information on the acoustic "startle" reflex.

*Jones, M. R. 1987. Dynamic pattern structure in music: recent theory and research. *Perception & Psychophysics* 41:621–634.

*Jones, M. R., and M. Boltz. 1989. Dynamic attending and responses to time. *Psychological Review* 96:459–491.
> Examples of Jones's work on timing and music.

*Keele, S. W., and R. Ivry. 1990. Does the cerebellum provide a common computation for diverse tasks—A timing hypothesis. *Annals of The New York Academy of Sciences* 608:179–211.
> Example of Ivry's work on timing and the cerebellum.

*Large, E. W., and M. R. Jones. 1995. The time course of recognition of novel melodies. *Perception and Psychophysics* 57 (2):136–149.

*———. 1999. The dynamics of attending: How people track time-varying events. *Psychological Review* 106 (1):119–159.
> More examples of Jones's work on timing and music.

*Lee, L. 2003. A report of the functional connectivity workshop, Düsseldorf 2002. *NeuroImage* 19:457–465.
> One of the papers Menon read to create the analyses we needed for our nucleus accumbens study.

*Levitin, D. J., and U. Bellugi. 1998. Musical abilities in individuals with Williams syndrome. *Music Perception* 15 (4):357–389.

*Levitin, D. J., K. Cole, M. Chiles, Z. Lai, A. Lincoln, and U. Bellugi. 2004. Characterizing the musical phenotype in individuals with Williams syndrome. *Child Neuropsychology* 10 (4):223–247.
    Information on Williams syndrome and two studies of their musical abilities.

*Levitin, D. J., and V. Menon. 2003. Musical structure is processed in "language" areas of the brain: A possible role for Brodmann Area 47 in temporal coherence. *NeuroImage* 20 (4):2142–2152.

*———. 2005. The neural locus of temporal structure and expectancies in music: Evidence from functional neuroimaging at 3 Tesla. *Music Perception* 22 (3):563–575.

*Levitin, D. J., V. Menon, J. E. Schmitt, S. Eliez, C. D. White, G. H. Glover, J. Kadis, J. R. Korenberg, U. Bellugi, and A. L. Reiss. 2003. Neural correlates of auditory perception in Williams syndrome: An fMRI study. *NeuroImage* 18 (1):74–82.
    Studies that showed cerebellar activations to music listening.

*Loeser, J. D., R. J. Lemire, and E. C. Alvord. 1972. Development of folia in human cerebellar vermis. *Anatomical Record* 173 (1):109–113.
    Background on cerebellar physiology.

*Menon, V., and D. J. Levitin. 2005. The rewards of music listening: Response and physiological connectivity of the mesolimbic system. *NeuroImage* 28 (1):175–184.
    The paper in which we showed the involvement of the nucleus accumbens and the brain's reward system in music listening.

*Merzenich, M. M., W. M. Jenkins, P. Johnston, C. Schreiner, S. L. Miller, and P. Tallal. 1996. Temporal processing deficits of language-learning impaired children ameliorated by training. *Science* 271:77–81.
    Paper showing that dyslexia may be caused by a timing deficit in children's auditory systems.

*Middleton, F. A., and P. L. Strick. 1994. Anatomical evidence for cerebellar and basal ganglia involvement in higher cognitive function. *Science* 266 (5184):458–461.

*Penhune, V. B., R. J. Zatorre, and A. C. Evans. 1998. Cerebellar contributions to motor timing: A PET study of auditory and visual rhythm reproduction. *Journal of Cognitive Neuroscience* 10 (6):752–765.

*Schmahmann, J. D. 1991. An emerging concept—the cerebellar contribution to higher function. *Archives of Neurology* 48 (11):1178–1187.

*Schmahmann, Jeremy D., ed. 1997. *The Cerebellum and Cognition*, International Review of Neurobiology, v. 41. San Diego: Academic Press.

*Schmahmann, S. D., and J. C. Sherman. 1988. The cerebellar cognitive affective syndrome. *Brain and Cognition* 121:561–579.
> Background information on the cerebellum, function, and anatomy.

*Tallal, P., S. L. Miller, G. Bedi, G. Byma, X. Wang, S. S. Nagarajan, C. Schreiner, W. M. Jenkins, and M. M. Merzenich. 1996. Language comprehension in language-learning impaired children improved with acoustically modified speech. *Science* 271:81–84.
> Paper showing that dyslexia may be caused by a timing deficit in children's auditory systems.

*Ullman, S. 1996. *High-level Vision: Object Recognition and Visual Cognition.* Cambridge: MIT Press.
> On the architecture of the visual system.

*Weinberger, N. M. 1999. Music and the auditory system. In *The Psychology of Music*, edited by D. Deutsch. San Diego: Academic Press.
> On the physiology and connectivity of the music/auditory system.

Chapter 7

*Abbie, A. A. 1934. The projection of the forebrain on the pons and cerebellum. *Proceedings of the Royal Society of London (Biological Sciences)* 115:504–522.
> Source of the quote about the cerebellum being involved in art.

*Chi, Michelene T. H., Robert Glaser, and Marshall J. Farr, eds. 1988. *The Nature of Expertise*. Hillsdale, N.J.: Lawrence Erlbaum Associates.
> Psychological studies of expertise, including chess players.

*Elbert, T., C. Pantev, C. Wienbruch, B. Rockstroh, and E. Taub. 1995. Increased cortical representation of the fingers of the left hand in string players. *Science* 270 (5234):305–307.
> Source for the cortical changes associated with playing violin.

*Ericsson, K. A., and J. Smith, eds. 1991. *Toward a General Theory of Expertise: Prospects and Limits*. New York: Cambridge University Press.
> Psychological studies of expertise, including chess players.

*Gobet, F., P. C. R. Lane, S. Croker, P. C. H. Cheng, G. Jones, I. Oliver, J. M. Pine. 2001. Chunking mechanisms in human learning. *Trends in Cognitive Sciences* 5:236–243.
> On chunking for memory.

*Hayes, J. R. 1985. Three problems in teaching general skills. In *Thinking and Learning Skills: Research and Open Questions*, edited by S. F. Chipman, J. W. Segal, and R. Glaser. Hillsdale, N.J.: Erlbaum.
> Source for the study that argued that Mozart's early works were not highly regarded, and refutation of the claim that Mozart didn't need ten thousand hours like everyone else to become an expert.

Howe, M. J. A., J. W. Davidson, and J. A. Sloboda. 1998. Innate talents: Reality or myth? *Behavioral & Brain Sciences* 21 (3):399–442.
    One of my favorite articles, although I don't agree with everything in it; an overview of the "talent is a myth" viewpoint.

Levitin, D. J. 1982. Unpublished conversation with Neil Young, Woodside, CA.

———. 1996. Interview: A Conversation with Joni Mitchell. *Grammy*, Spring, 26–32.

———. 1996. Stevie Wonder: Conversation in the Key of Life. *Grammy*, Summer, 14–25.

———. 1998. Still Creative After All These Years: A Conversation with Paul Simon. *Grammy*, February, 16–19, 46.

———. 2000. A conversation with Joni Mitchell. In *The Joni Mitchell Companion: Four Decades of Commentary*, edited by S. Luftig. New York: Schirmer Books.

———. 2001. Paul Simon: The Grammy Interview. *Grammy*, September, 42–46.

———. 2004. Unpublished conversation with Joni Mitchell, December, Los Angeles, CA.
    Sources for the anecdotes and quotations from these musicians about musical expertise.

MacArthur, P. (1999). JazzHouston Web site. http:www.jazzhouston.com/forum/messages.jsp?key=352&page=7&pKey=1&fpage=1&total=588.
    Source of the quote about Rubinstein's mistakes.

*Sloboda, J. A. 1991. Musical expertise. In *Toward a General Theory of Expertise*, edited by K. A. Ericcson and J. Smith. New York: Cambridge University Press.
    Overview of issues and findings in musical expertise literature.

Tellegen, Auke, David Lykken, Thomas Bouchard, Kimerly Wilcox, Nancy Segal, and Stephen Rich. 1988. Personality similarity in twins reared apart and together. *Journal of Personality and Social Psychology* 54 (6):1031–1039.
    The Minnesota Twins study.

*Vines, B. W., C. Krumhansl, M. M. Wanderley, and D. Levitin. In press. Cross-modal interactions in the perception of musical performance. *Cognition*.
    Source of the study about musician gestures conveying emotion.

Chapter 8

*Berlyne, D. E. 1971. *Aesthetics and Psychobiology*. New York: Appleton-Century-Crofts.
    On the "inverted-U" hypothesis of musical liking.

*Gaser, C., and G. Schlaug. 2003. Gray matter differences between musicians and nonmusicians. *Annals of the New York Academy of Sciences* 999:514–517.
    Differences between the brains of musicians and nonmusicians.

*Husain, G., W. F. Thompson, and E. G. Schellenberg. 2002. Effects of musical tempo and mode on arousal, mood, and spatial abilities. *Music Perception* 20 (2):151–171.
    The "Mozart Effect" explained.

*Hutchinson, S., L. H. Lee, N. Gaab, and G. Schlaug. 2003. Cerebellar volume of musicians. *Cerebral Cortex* 13:943–949.
    Differences between the brains of musicians and nonmusicians.

*Lamont, A. M. 2001. Infants' preferences for familiar and unfamiliar music: A socio-cultural study. Paper read at Society for Music Perception and Cognition, August 9, 2001, at Kingston, Ont.
    On infants' prenatal musical experience.

*Lee, D. J., Y. Chen, and G. Schlaug. 2003. Corpus callosum: musician and gender effects. *NeuroReport* 14:205–209.
    Differences between the brains of musicians and nonmusicians.

*Rauscher, F. H., G. L. Shaw, and K. N. Ky. 1993. Music and spatial task performance. *Nature* 365:611.
    The original report of the "Mozart Effect."

*Saffran, J. R. 2003. Absolute pitch in infancy and adulthood: the role of tonal structure. *Developmental Science* 6 (1):35–47.
    On the use of absolute pitch cues by infants.

*Schellenberg, E. G. 2003. Does exposure to music have beneficial side effects? In *The Cognitive Neuroscience of Music*, edited by I. Peretz and R. J. Zatorre. New York: Oxford University Press.

*Thompson, W. F., E. G. Schellenberg, and G. Husain. 2001. Arousal, mood, and the Mozart Effect. *Psychological Science* 12 (3):248–251.
    The "Mozart Effect" explained.

*Trainor, L. J., L. Wu, and C. D. Tsang. 2004. Long-term memory for music: Infants remember tempo and timbre. *Developmental Science* 7 (3):289–296.
    On the use of absolute-pitch cues by infants.

*Trehub, S. E. 2003. The developmental origins of musicality. *Nature Neuroscience* 6 (7):669–673.

*———. 2003. Musical predispositions in infancy. In *The Cognitive Neuroscience of Music*, edited by I. Peretz and R. J. Zatorre. Oxford: Oxford University Press.
    On early infant musical experience.

Chapter 9

Barrow, J. D. 1995. *The Artful Universe*. Oxford, UK: Clarendon Press.
    "Music has no role in survival of the species."

Blacking, J. 1995. *Music, Culture, and Experience*. Chicago: University of Chicago Press.

"The embodied nature of music, the indivisibility of movement and sound, characterizes music across cultures and across time."

Buss, D. M., M. G. Haselton, T. K. Shackelford, A. L. Bleske, and J. C. Wakefield. 2002. Adaptations, exaptations, and spandrels. In *Foundations of Cognitive Psychology: Core Readings*, edited by D. J. Levitin. Cambridge: MIT Press.

I've intentionally avoided making a distinction between two types of evolutionary by-products, *spandrels* and *exaptations*, in order to simplify the presentation in this chapter, and I've used the term *spandrels* for both types of evolutionary by-products. Because Gould himself did not use the terms consistently through his writings, and because the main point is not compromised by glossing over this distinction, I present a simplified explanation here, and I don't think that readers will suffer any loss of understanding. Buss, et al., discuss this distinction and others, based on the work of Stephen Jay Gould cited below.

*Cosmides, L. 1989. The logic of social exchange: Has natural selection shaped how humans reason? *Cognition* 31:187–276.

*Cosmides, L., and J. Tooby. 1989. Evolutionary psychology and the generation of culture, Part II. Case Study: A computational theory of social exchange. *Ethology and Sociobiology* 10:51–97.

Perspectives of evolutionary psychology on cognition as adaptation.

Cross, I. 2001. Music, cognition, culture, and evolution. *Annals of the New York Academy of Sciences* 930:28–42.

———. 2001. Music, mind and evolution. *Psychology of Music* 29 (1):95–102.

———. 2003. Music and biocultural evolution. In *The Cultural Study of Music: A Critical Introduction*, edited by M. Clayton, T. Herbert and R. Middleton. New York: Routledge.

———. 2003. Music and evolution: Consequences and causes. *Comparative Music Review* 22 (3):79–89.

———. 2004. Music and meaning, ambiguity and evolution. In *Musical Communications*, edited by D. Miell, R. MacDonald and D. Hargraves.

The sources for Cross's arguments as articulated in this chapter.

Darwin, C. 1871/2004. *The Descent of Man and Selection in Relation to Sex*. New York: Penguin Classics.

The source for the ideas Darwin had about music, sexual selection, and adaptation. "I conclude that musical notes and rhythm were first acquired by the male or female progenitors of mankind for the sake of charming the opposite sex. Thus musical tones became firmly associated with some of the strongest passions an animal is capable of feeling, and are consequently used instinctively. . . ."

*Deaner, R. O., and C. L. Nunn. 1999. How quickly do brains catch up with bodies? A comparative method for detecting evolutionary lag. *Proceedings of the Royal Society of London B* 266 (1420):687–694.
On evolutionary lag.

Gleason, J. B. 2004. *The Development of Language*, 6th ed. Boston: Allyn & Bacon.
Undergraduate text on the development of language ability.

*Gould, S. J. 1991. Exaptation: A crucial tool for evolutionary psychology. *Journal of Social Issues* 47:43–65.
Gould's explication of different kinds of evolutionary by-products.

Huron, D. 2001. Is music an evolutionary adaptation? In *Biological Foundations of Music*.
Huron's response to Pinker (1997); the idea of comparing autism to Williams syndrome for an argument about the link between musicality and sociability first appeared here.

*Miller, G. F. 1999. Sexual selection for cultural displays. In *The Evolution of Culture*, edited by R. Dunbar, C. Knight and C. Power. Edinburgh: Edinburgh University Press.

*———. 2000. Evolution of human music through sexual selection. In *The Origins of Music*, edited by N. L. Wallin, B. Merker and S. Brown. Cambridge: MIT Press.

———. 2001. Aesthetic fitness: How sexual selection shaped artistic virtuosity as a fitness indicator and aesthetic preferences as mate choice criteria. *Bulletin of Psychology and the Arts* 2 (1):20–25.

*Miller, G. F., and M. G. Haselton. In Press. Women's fertility across the cycle increases the short-term attractiveness of creative intelligence compared to wealth. *Human Nature*.
Source articles for Miller's view on music as sexual fitness display.

Pinker, S. 1997. *How the Mind Works*. New York: W. W. Norton.
Source of Pinker's "auditory cheesecake" analogy.

Sapolsky, R. M. *Why Zebras Don't Get Ulcers*, 3rd ed. 1998. New York: Henry Holt and Company.
On evolutionary lag.

Sperber, D. 1996. *Explaining Culture*. Oxford, UK: Blackwell.
Music as an evolutionary parasite.

*Tooby, J., and L. Cosmides. 2002. Toward mapping the evolved functional organization of mind and brain. In *Foundations of Cognitive Psychology*, edited by D. J. Levitin. Cambridge: MIT Press.
Another work by these evolutionary psychologists on cognition as adaptation.

Turk, I. *Mousterian Bone Flute*. Znanstvenoraziskovalni Center Sazu 1997 [cited December 1, 2005. Available from http:www.uvi.si/eng/slovenia/background-information/neanderthal-flute/.]

The original report on the discovery of the Slovenian bone flute.

*Wallin, N. L. 1991. *Biomusicology: Neurophysiological, Neuropsychological, and Evolutionary Perspectives on the Origins and Purposes of Music*. Stuyvesant, N.Y.: Pendragon Press.

*Wallin, N. L., B. Merker, and S. Brown, eds. 2001. *The Origins of Music*. Cambridge: MIT Press.

Further reading on the evolutionary origins of music.

ACKNOWLEDGMENTS

I would like to thank all the people who helped me to learn what I know about music and the brain. For teaching me how to make records, I am indebted to the engineers Leslie Ann Jones, Ken Kessie, Maureen Droney, Wayne Lewis, Jeffrey Norman, Bob Misbach, Mark Needham, Paul Mandl, Ricky Sanchez, Fred Catero, Dave Frazer, Oliver di Cicco, Stacey Baird, Marc Senasac, and the producers Narada Michael Walden, Sandy Pearlman, and Randy Jackson; and for giving me the chance to, Howie Klein, Seymour Stein, Michelle Zarin, David Rubinson, Brian Rohan, Susan Skaggs, Dave Wellhausen, Norm Kerner, and Joel Jaffe. For their musical inspiration and time spent in conversation I am grateful to Stevie Wonder, Paul Simon, John Fogerty, Lindsey Buckingham, Carlos Santana, kd lang, George Martin, Geoff Emerick, Mitchell Froom, Phil Ramone, Roger Nichols, George Massenburg, Cher, Linda Ronstadt, Peter Asher, Julia Fordham, Rodney Crowell, Rosanne Cash, Guy Clark, and Donald Fagen. For teaching me about cognitive psychology and neuroscience, Susan Carey, Roger Shepard, Mike Posner, Doug Hintzman, and Helen Neville. I am grateful to my collaborators, Ursula Bellugi and Vinod Menon, who have given me an exciting and rewarding second career as a scientist, and to my close colleagues Steve McAdams, Evan Balaban, Perry Cook, Bill Thompson, and Lew Goldberg. My students and postdoctoral fellows have been an additional source of pride and inspiration, and helped with their comments on drafts of this book: Bradley Vines, Catherine Guastavino, Susan Rogers, Anjali Bhatara, Theo Koulis, Eve-Marie Quintin, Ioana Dalca, Anna Tirovolas, and Andrew Schaaf. Jeff Mogil, Evan Balaban, Vinod Menon, and Len Blum provided valuable comments on portions of the manuscript. Still, any errors are my own. My dear friends Michael Brook and Jeff Kimball have helped me throughout the writing of this book in many ways, with their conversation, questions, support, and musical insights. My department chair, Keith Franklin, and the dean of the Schulich School of Music, Don McLean, have provided me with an enviably productive and supportive intellectual environment within which to work.

I would also like to thank my editor at Dutton, Jeff Galas, for his guidance and support through every step of turning these ideas into a book, for his hundreds of suggestions and excellent advice, and Stephen Morrow at Dutton for his helpful contributions in editing the manuscript; without Jeff and Stephen, this book would not have existed. Thank you both.

The subtitle for Chapter 3 is taken from the excellent book edited by R. Steinberg and published by Springer-Verlag.

And thank you to my favorite pieces of music: Beethoven's Sixth Symphony; "Joanne" by Michael Nesmith; "Sweet Georgia Brown" by Chet Atkins and Lenny Breau; and "The End" by the Beatles.

INDEX

Note: Page numbers in *italics* refer to illustrations or charts.